Plastic Footprint

Pankaj Pathak · Sadia Ilyas · Rajiv R. Srivastava
Editors

Plastic Footprint

Global Issues, Impacts and Solutions

Springer

Editors
Pankaj Pathak
Resource Management Lab
Department of Environmental Science and Engineering
SRM University-AP
Amaravati, Andhra Pradesh, India

Rajiv R. Srivastava
Process Metallurgy
Department of Civil
Environmental and Natural Resources Engineering
Wallenberg Initiative Materials Science for Sustainability (WISE)
Luleå University of Technology
Luleå, Sweden

Sadia Ilyas
Process Metallurgy
Minerals and Metallurgical Engineering Division
Department of Civil
Environmental and Natural Resources Engineering
Wallenberg Initiative Materials Science for Sustainability (WISE)
Luleå University of Technology
Luleå, Sweden

ISBN 978-981-96-6361-3 ISBN 978-981-96-6362-0 (eBook)
https://doi.org/10.1007/978-981-96-6362-0

This Springer imprint is published by the registered company Springer Nature Singapore Pte Ltd.
The registered company address is: 152 Beach Road, #21-01/04 Gateway East, Singapore 189721, Singapore

Preface

Plastic pollution has become mammoth issue worldwide, with plastic waste generation far outpacing than its degradation. Around 400 million tons of plastic waste are generated annually, causing significant harm to both human health and the environment. Plastic trash which is disposed of on land and ocean gets converted to micro- and nano-plastics and shows a significant contribution to climate change. Apart from this, plastics enter the food chain, and a substantial volume of plastics is found in the human body. Sustainable plastic management, including converting plastic waste into fuel, offers a potential solution that addresses both pollution and the global energy demand. This book explores the myths and benefits of plastics, the global scale of plastic pollution, and recycling efforts worldwide. It also examines how transforming plastic waste into fuel can provide a sustainable resource for the twenty-first century.

This book is a valuable resource for scholars, researchers, and academics focused on addressing plastic pollution and finding viable solutions. Key highlights include an assessment of plastic pollution, the pros and cons of plastics, emerging contaminants like micro- and nano-plastics, and strategies for their remediation to ensure a sustainable environment for modern society. The first chapter "Plastic Utilization in the Modern Era: The Twenty-First Century" highlights the utilization of plastics and their potential impacts in the twenty-first century. The second chapter "Micro- and Nano-Plastics (MNPs) in the Environment: Addressing the Sampling and Identification Issues" focuses on emerging contaminants, specifically micro- and nano-plastics, and their identification and collection in the environment. The third chapter "Advancing Sustainability in Construction: Integrating Recycled Plastic Waste into Green Concrete" addresses the sustainable use of plastics in construction materials. The fourth and fifth chapters "Plastic Pollution Disrupting the Aquatic Ecosystem" and "Development of a Standardized Plastic Footprint Measurement Tool for Combating Marine Plastic Pollution" examine the impact of plastics in aquatic environments. The sixth chapter "Nanobiotechnology for Reducing Plastic Footprint and Combatting Micro and Nano-plastic Pollution" highlights the role of nanomaterials in mitigating plastic pollution. The seventh chapter "Assessing Effects and Impacts Associated with Macro–Micro Plastics in Waterbodies" assesses the environmental impact of micro- and nano-plastics. The eighth chapter “Assessing Different Tools Employed in Auto-segregation of Plastic Waste” discusses the use of artificial intelligence to auto-segregate the plastics.

Nonetheless, the material collected in this volume offers comprehensive insights into the plastic footprint, its environmental impacts, and the mitigation efforts undertaken in recent years to foster sustainability within the plastic industry. Overall, the content in this book brings in-depth understanding and extension of knowledge in the field of plastic pollution and its associated impacts. Dr. Pankaj Pathak, Dr. Sadia Ilyas, and Dr. Rajiv R. Srivastava individually acknowledge the authors and reviewers for their invaluable contributions, expertise, and dedicated efforts in shaping this book. Additionally, Dr. Ilyas greatly acknowledges her participation through the Wallenberg Initiative Materials Science for Sustainability (WISE) funded by the Knut and Alice Wallenberg Foundation.

Amaravati, India — Dr. Pankaj Pathak
Luleå, Sweden — Dr. Sadia Ilyas
Luleå, Sweden — Dr. Rajiv R. Srivastava

Contents

About the Editors

Dr. Pankaj Pathak is an Associate Professor in the Department of Environmental Science and Engineering at SRM University—Andhra Pradesh, India. Her research focuses on solid and hazardous waste management, with a particular emphasis on plastic and e-waste recycling, as well as waste-to-resource strategies. She has keen interests in circular economy principles, life-cycle assessment, techno-economic analysis, and the sustainable recycling of green energy resources. She has published numerous peer-reviewed articles in prestigious international journals and contributed to books published by Springer, CRC Press, and other renowned academic publishers.

Dr. Sadia Ilyas is an Associate Professor in Process Metallurgy and a Fellow of the Wallenberg Initiative Materials Science for Sustainability (WISE) at the Department of Civil, Environmental and Natural Resources Engineering, Luleå University of Technology in Sweden. Previously, she has served as a Brainpool Scientist under the National Research Foundation of Korea and held various academic and research positions in South Korea, Pakistan, and China. Her research domain is mainly focused on sustainable technological advancements in natural resources management, urban mining, circular economy, green chemistry, and impact assessment using life-cycle and techno-economic analysis tools. She is among top 2% of scientists in her field, and published several highly cited articles in different peer-reviewed journals, edited books, and contributed chapters. She holds several patents, and also actively involved in the editorial roles at several high-indexed journals.

Dr. Rajiv R. Srivastava is an academician cum industrial researcher specializing in waste management and sustainable environmental technology, with a focus on hydrometallurgical processes and reaction thermodynamics. He is a WISE Researcher at the Department of Civil, Environmental and Natural Resources Engineering, Luleå University of Technology in Sweden. In addition to his academic work, he has made significant contributions to industrial R&D, particularly in the commercialization of metal recycling processes, including the recovery of metals from spent Li-ion batteries, spent HDS catalysts, and e-waste. Over the course of his career, he has collaborated with various R&D institutions, universities, and industries in multiple roles. His interdisciplinary research interests span solid and industrial waste management, urban mining, and the (hydro)metallurgical recovery of energy-critical elements for clean and alternative energy production. He has published numerous research articles, holds patents, and is ranked among the top 2% of scientists in his field.

Plastic Utilization in the Modern Era: The Twenty-First Century

Hephzi Jones and Pankaj Pathak

Abstract Plastics are among the greatest industrial innovations in the world, but their extensive use and improper disposal methods are making them more and more harmful to the environment and human health. Marine pollution, chemical contamination, biodiversity loss, and climate change are some of these repercussions. This calls for immediate action to address these pressing issues. Due to the enormous demand and frequent use, several nations are now affected by plastic waste that requires rapid attention and effective management strategies. Polymer pollution on terrestrial and aquatic is on the rise worldwide, highlighting the far-reaching consequences of plastic solid waste on both living and non-living elements of the ecosystem. Accordingly, in this chapter, we focus on the various applications of plastics, their disposal rates, and their impacts on the human environment. Furthermore, this chapter critically evaluates and suggests circular economy solutions to mitigate these impacts and contribute to achieving the sustainable development goals.

Keywords Plastic · Microplastics · Environmental impact · Recycling · Circular economy

1 Introduction

Plastics, invented over a century ago, have transformed our lifestyles. This extraordinary material has revolutionized every aspect of human activity and is widely employed in various applications, including films and sheets, fibers, yarns, adhesives, coatings, and particles. Industry sectors include electronics, agriculture, biomedical devices, construction, textiles, home, automotive, leisure, sports, and packaging.

H. Jones · P. Pathak (✉)
Resource Management Lab, Department of Environmental Science and Engineering, SRM University AP, Amaravati, Andhra Pradesh, India
e-mail: pankajpathak18@gmail.com

P. Pathak
Waste to Wealth Group, SRM University AP, Amaravati, Andhra Pradesh, India

P. Pathak et al. (eds.), *Plastic Footprint*, https://doi.org/10.1007/978-981-96-6362-0_1

These polymers are classified as thermosets, thermoplastics, elastomers, or thermoplastic elastomers (Millican & Agarwal, 2021). Humans have experienced various eras, some named after the materials used to craft significant tools and necessities, such as the Stone Age, Bronze Age, and Iron Age. Today's era, often called the Plastic Age, reflects a similar pattern. The term "plastic" is derived from the Greek word "plastikos, 'meaning' capable of being molded or shaped." Renowned for its adaptability and usefulness, plastic is a synthetic polymer that combines various compounds with monomers derived from petrochemicals (Torres-Agullo et al., 2021).

The world's plastics business started in 1907, and polymer was invented in 1860. However, the 1920s saw a major expansion of industry, which made it possible to produce plastic on a huge scale. The history of plastics, illustrated in Fig. 1, reveals substantial growth during the 1940s. Plastic production soared from 2 million metric tons in 1950 to 368 million metric tons in 2019 and between 1950 and 2018, plastic usage increased nearly 180-fold. In 2022, global plastic production was estimated to reach around 400.3 million tons. Projections suggest that polymer production will continue to grow significantly, reinforcing its importance as a key material of the twenty-first century (Hassanpour & Unnisa, 2017). Over one-third of this rate is attributed to using plastic bags for packaging, while another third is allocated to housing items such as plastic pipes. Additionally, 10% of vinyl cladding ends up in the ocean. According to estimates, 60–80% of marine garbage comprises plastics. Of this plastic litter, 80% originates from terrestrial sources, while 18% is linked to the fishing sector (Nayanathara Thathsarani Pilapitiya & Ratnayake, 2024).

The use of plastics, often mistakenly regarded as "a contemporary symbol of modernity," has spiraled out of control, becoming an undeniable symbol of the current crisis in post-modern civilization. Plastics, deeply embedded in everyday life,

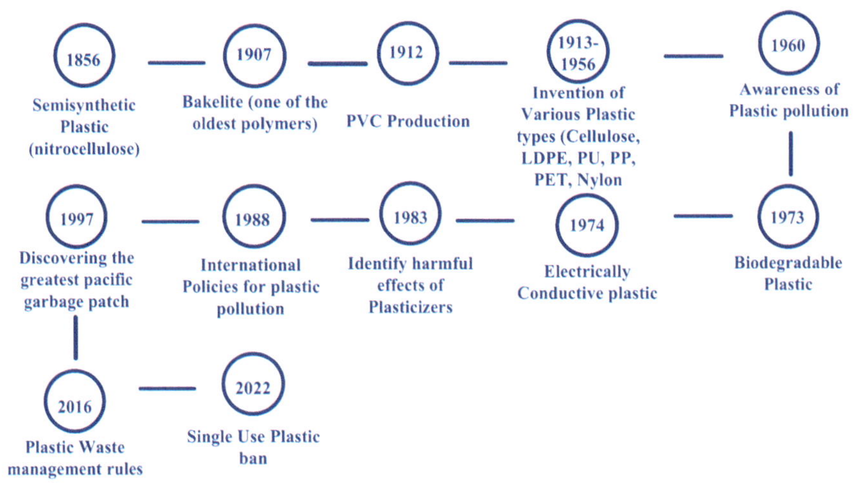

Fig. 1 Chronology of plastic from 1856 to 2022

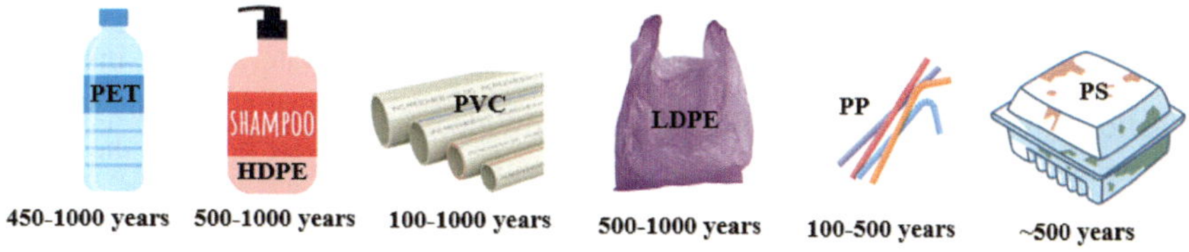

Fig. 2 Degradation life span of different types of polymers

have dramatically transformed the modern world. However, their persistent nature as a material that "does not go away" has made them a significant environmental challenge (Haram et al., 2020). In comparison to other wastes, plastics take over a millennium to decompose and Fig. 2 shows how long it takes for various plastic types to decompose (Babaremu et al., 2022). Massive environmental degradation has resulted from manufacturing and dumping plastics, endangering marine life and compromising the food chain. Concerns regarding the effects of microplastics on human health have been raised by the discovery of these particles in human blood, lungs, and guts. Population increases, urbanization, and economic development are the primary causes of the rise of plastic waste (Amasuomo & Baird, 2016; Nandikes et al., 2024; Zhang et al., 2020). Humans are at risk from the material, which also affects tourism and leisure; endangers rivers, coastlines, and marine life; and is an artistic issue that exposes important and disagreeable elements of human nature (Corraini et al., 2018).

Approximately 400.3 million metric tons of plastic are produced worldwide each year, with 79% ending up in landfills, 12% being incinerated, and only 9% recycled (Jones & Pathak, 2024; Tejaswini et al., 2022). Consequently, plastic waste is accumulating at a startling rate, and the world feels ashamed of its failure to solve the issue. Several initiatives to combat plastic pollution have concentrated on cleanup and waste management. Others have stressed reducing and getting rid of plastic. The existing linear "take, make, use, dispose of" economic paradigm needs to be replaced with a thorough Circular Economy (CE) strategy (Anwar et al., 2021; Pujara et al., 2023). The CE approach provides a way to reuse, conserve, and repurpose products and materials in the most economical, ecologically, functionally, and socially responsible ways. The main three principles of the CE strategy are.

(i) Improving product design to minimize waste and pollution.
(ii) Preserving existing sources and products.
(iii) Regenerating and conserving natural systems.

This results in a continuous development cycle that extends product life, reduces greenhouse gas emissions, and conserves limited raw material resources (Ahmed, 2023). By implementing the circular economy concept, which recycles plastics rather than discarding them, we may recycle plastics back into the manufacturing process to create new goods, increasing the value of plastic waste as a feedstock and creating chances for further expansion. The circular economy sees and treats plastics as a feedstock for numerous new products, extending their potential beyond "make-use-dispose."

2 Plastic Applications in Different Sectors

Plastics provide numerous benefits, including low electrical and thermal conductivity, lightweight characteristics, high strength, toughness, stiffness, versatility, design flexibility, insulating properties, and corrosion resistance. The low production cost, malleability, and ease of processing of plastic have contributed to its extensive use in everyday life (Ilyas et al., 2018). These qualities have led to increased use of plastics, particularly during World War II when nylon was invented. Polymers such as PET, PE, PVC, PP, and PS are widely utilized in consumer and non-packaging plastic items. Polymers are colored, shaped, and textured with various dyes and chemicals. PE, the most widely used polymer, comes in three types: HDPE, LDPE, and LLDPE. It is commonly used in household items including plastic containers, bottles, bags, and toys. Figure 3 depicts the estimated consumption share of polymers for several plastic product types in 2020. PP, along with the three forms of PE, accounted for more than half of India's polymer use in plastic products in 2018–19 (Circular Economy for Plastics in India: A roadmap, 2021).

Plastics have various applications, including agriculture, building, heat insulation, transportation, manufacturing, packing, furniture, electronics, leisure goods, vehicles, and medicine. Due to their lightweight nature and biocompatibility, plastics are used in 85% of medical devices, including intravenous bags, disposable syringes, sterile packaging, and joint replacements. Plastics provide several advantages, such as.

(i) Environmental benefits (e.g., PET beverage packaging reduces energy consumption and greenhouse gas emissions, while lighter plastic composites in aircraft decrease fuel usage and streamline assembly).
(ii) Social benefits (e.g., using a variety of water distribution and control systems to supply and store clean drinking water, and using food packaging to store fresh goods in a safe, time-dependent manner with regulated temperature and pressure).

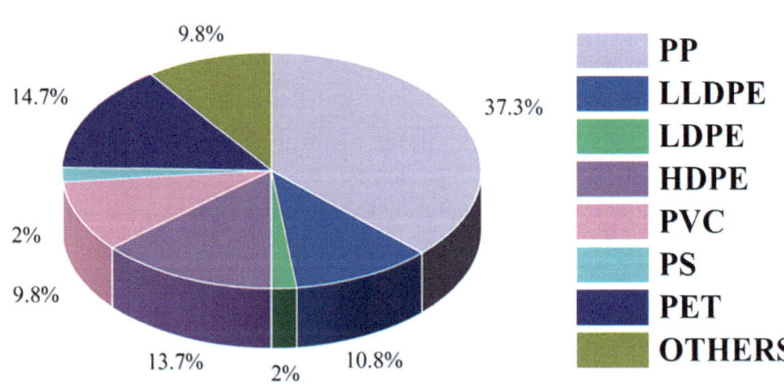

Fig. 3 Consumption shares of polymers

(iii) Economic benefits (e.g., using inexpensive plastic materials to replace automobile parts) are shown in Table1.

3 Disposal Rate of Plastics

Conventional plastics account for around 9% of worldwide oil and gas output, 5% as feedstock, and roughly 4% of the energy used in production. Currently, 74% of Europe's post-consumer plastic garbage is burnt or sent to dump sites. Except for numerous long-term-use plastics such as water pipes, electric cables, household items, and electronics casing, the majority of consumer plastic commodities are designed to be disposable after one use or have a short usage life. Table 2 presents post-consumer plastic waste rates by country around the world (Hassanpour & Unissa, 2017; Shankar et al., 2023).

Post-consumer plastic trash in Europe amounts to about 25 million tons annually, of which 6.6 million are recovered primarily by mechanical recycling, 8.9 million are used to meet energy conversion targets, and 9.6 million are dumped in landfills. In Canada, almost 2.8 million tons of plastic trash are disposed of annually (Hassanpour and Unissa, 2017). Like all consumer goods, plastic waste is disposed of in a way that increases municipal waste and creates litter on the streets. The number of abandoned materials, notably thermoplastic items, found in urban waste is growing. Plastic waste can linger in the environment for a long time, and thermoplastics are not easily biodegradable. The exact lifespan of discarded plastics differs significantly based on the material's chemical composition, the environmental conditions it is exposed to, and the definition or measurement of "degradation." While compostable plastics can break down within months, conventional plastics take much longer. Decomposition may result from biological or environmental factors such as wind, rain, and sunlight. Plastics are extremely persistent in the environment due to their chemical resistance, which makes it difficult to pinpoint the origins of plastic litter when compared to other discarded materials like lignocellulosic paper. However, as littering is a behavioral problem, education is the main way to combat it. The best way to combat littering is to raise consumer knowledge of its negative effects on the environment, which requires more time and money (Andrady & Neal, 2009).

4 Environmental and Human Impacts of Plastic Waste

The unplanned dumping and disposal of plastic litter is the source of the problem. It damages the natural environment by generating problems for plants, wildlife, and the human population, as well as the area's aesthetics and cleanliness (Carbonel et al., 2023). The sewer network and drain route are clogged by plastic garbage causing flooding, mosquito breeding grounds, disease-carrying flies, and rodents (Verma et al., 2016). They harm aquatic bodies by releasing a lot of pollutants into them.

Table 1 Different grades of plastic production, generation of waste, and treatment technologies (Evode et al., 2021; Karmakar, 2022; Kibria et al., 2023; Plastics Europe, 2022; Shah et al., 2023; Our World in Data, 2019; Shah et al., 2023)

Plastic type	Name of plastic	Application	Generation (2019)	Recyclable/ non-recyclable	Plastic waste generation	Treatment technologies
PET	Polyethylene terephthalate	Water bottles, jelly pickles, soft drink bottles, food jars, plastic films, sheets, salad dressing containers, processed meat packages, etc	24.92 million tons	Recyclable	24.81 million tons	Methanolysis, hydrolysis, glycolysis, aminolysis, ammonolysis, incineration, pyrolysis, gasification
HDPE	High-density polyethylene	Food containers, plastic bags, bottles, woven sacks, detergent bags, metalized pouches, milk pouches & shopping bags, agricultural pipe, playground equipment, recycling bins, freezer bags, lids, soap containers, faux-wood planks, shampoo bottles, protective helmets, recycled wood-plastic composites, vehicle fuel tanks, insulation, plastic toys, and more	55.54 million tons	Recyclable	44.71 million tons	Pyrolysis incineration gasification
PVC	Polyvinyl chloride	Cable insulations, pipes, hoses, sheets, floorings, wire, multilayer tubes, plumbing, door and window profiles, wiring, electrical cable insulation, electric conductors, cleaning solution containers, clothing, medical containers, signage, cladding, vinyl records, fruit plastic packing, sweet trays, blister packaging, etc	51.39 million tons	Recyclable	21.2 million tons	Incineration

(continued)

Table 1 (continued)

Plastic type	Name of plastic	Application	Generation (2019)	Recyclable/ non-recyclable	Plastic waste generation	Treatment technologies
LDPE	Low-density polyethylene	Plastic bags, food containers, bottles, plastic toys, woven sacks, milk pouches and shopping bags, metalized pouches, films, six-pack rings, dispensing bottles, drink cartons, bin garbage, machine parts, laundry bags, protective shells, computer hardwires, trays, bin bags, playground fixtures, shrink wraps, etc	54.30 million tons	Recyclable	49.21 million tons	Pyrolysis incineration gasification
PP	Polypropylene	Disposable cups, bottle caps, straws, medicine bottles, cereal liners, packaging films, tupperware, car parts, thermal vests, disposable diapers, crisp bags, yogurt containers, hangers, funnels, trays, packaging tape, lunch boxes, clothing, etc	72.81 million tons	Recyclable	62 million tons	Pyrolysis incineration gasification
PS	Polystyrene	Disposable cups, CD covers, foams, glasses, plates, spoons, Cassette boxes, trays, beverage cups, dinnerware, melamine plates, insulation, packing material to egg cartons and disposable takeaway and hard packaging toys, etc	21.12 million tons	Recyclable	15.17 million tons	Incineration gasification
OTHER	Thermoset, polycarbonate (PC), polyurethane (PU) FRP	CD, helmets, PUF Bakelite, shoe soles, multilayer and laminated plastics, Nylon, lenses for sunglasses, sports, and safety goggles, mobile phones, baby bottles, coatings, foams, adhesives, paints, upholstery, insulators, etc	80.96 million tons	Not recyclable	57.56 million tons	Pyrolysis gasification

(continued)

Table 1 (continued)

Plastic type	Name of plastic	Application	Generation (2019)	Recyclable/ non-recyclable	Plastic waste generation	Treatment technologies
SUP	HDPE, LDPE, PET, PS, PP (or), EPS	Carry bags, non-woven carry bags, straws/ stirrers, small wrapping/packaging film, cutlery, film, Disposable rigid cups, cigarette overwrap, trays, and containers, plastic sticks (for balloons, flags, candies, etc.), EPS (thermocol) for decoration, plastic banners, cigarette overwrap film, disposable rigid cups, trays, and containers, wrapping film for food applications, multi-layered packaging	20 million tons (Indian Production)	–	5.5 million tons (India SUP)	–

Table 2 Post-consumer plastic waste around the world (Hassanpour & Unissa, 2017; Shankar et al., 2023)

Region	Energy recovery rate (%)	Recycling (%)	Disposal (%)
India	5	94.17	–
Switzerland	76	24	–
Norway	55	37	8
Estonia	26	32	46
Austria	72	24	4
Finland	44	18	38
Denmark	66	28	6
Belgium	62	32	6
Germany	63	33	4
Sweden	61	34	5
Luxemburg	70	24	6
Netherlands	59	33	8
Portugal	13	34	53
Spain	16	28	56
Ireland	25	31	44
Great Britain	9	22	69

When plastic is consumed by animals, it can cause choking in the stomach. Additionally, they reduce soil fertility and release volatile organic carbon when disposed of in landfills. Because they prevent the soil from absorbing water and nutrients, they slow down plant growth (Gupta, 2021). Microplastics in soils, sediments, and waters could have long-term detrimental effects on terrestrial ecosystems worldwide by negatively impacting organisms that are crucial to essential ecosystem services and processes, such as fungi and soil-dwelling invertebrates. In agricultural soils, organic fertilizers have been found to contain up to 895 microplastic particles per kilogram. Each year, up to 730,000 tons of microplastics are transported to agricultural areas in North America and Europe from urban sewage sludges used as manure. These microplastics may have direct impacts on crops, livestock, and soil ecosystems, or they may carry harmful chemicals (Plastics and the Circular Economy, 2018).

At present, the world produces a vast amount of plastic waste, as illustrated in Table 1, due to insufficient management, inefficient waste management systems, and ongoing contamination (Nayanathara Thathsarani Pilapitiya & Ratnayake, 2024). Plastic pollution can exceed a critical threshold, leading to irreversible global effects on biodiversity, ecosystems, and climate. This is due to the life cycle of plastics now posing a threat to planetary boundaries. Each year, roughly 14 million tons of plastic waste enter the ocean, predominantly from coastal areas in Asia. Over time, an estimated 700 aquatic species have experienced the deadly consequences of plastic pollution (Babaremu et al., 2022).

Accordingly, by 2050, it is projected that almost every seabird species worldwide will start consuming plastic waste. Most plastic debris accumulating in oceanic areas has been discovered to harbor harmful species of viruses, algae, and microbial communities, which readily carry toxic substances that ultimately disrupt genetic diversity and transform ecosystems (Villarrubia Gomez et al., 2018) if preventive or damage control measures are not implemented. In that case, there is a chance that the number of plastics in the ocean will increase to an estimated 30 million tons per year by 2040, which would harm the environment (Kehinde et al., 2020). On the other hand, when plastics are openly burned, many compounds responsible for their unique properties—such as hardness, durability, malleability, color, and plasticity—release toxic gaseous pollutants, including carcinogenic dioxins and furans, into the environment. These emissions contain various airborne toxins that damage human skin and eyes; increase the risk of cancer, heart disease, ulcers, birth defects, respiratory illnesses, thyroid problems, headaches, and nausea; and may even impair reproductive and neurological systems. The 2014 UN Environment report estimated the natural capital cost of plastics, including health impacts, climate change, and environmental degradation, at approximately USD 75 billion annually, with 75% of these environmental costs arising during the production stage, while a more recent study suggests that the environmental cost could reach USD 139 billion (Plastics and the Circular Economy, 2018).

5 Strategies for Implementing a Circular Economy in the Plastic Industry

The goals of a circular economy in the plastics industry focus on enhancing the economic feasibility of recycling and reusing plastics, reducing plastic leakage into the environment—particularly into waterways and oceans—and decreasing reliance on fossil-fuel-based feedstocks for plastic production by shifting to renewable alternatives (Plastics and the Circular Economy, 2018). Recent scientific and technological advancements provide examples of potential methods to accomplish these goals.

(i) **Manufacture Plastics from Alternative Raw Materials**.

Alternative feedstocks include CO_2, methane, oils, starch, cellulose, biopolymers, sewage sludge, and food items. Some plastics can be made from safe and biodegradable resources. Eco-friendly alternatives to flame retardants have been developed to reduce the usage of harmful chemicals in plastic manufacturing. Increased manufacturing of plastics from renewable sources can greatly reduce reliance on fossil fuels and minimize the negative consequences of the existing linear production and consumption model. Redesigning production processes and goods can enhance durability, reusability, and recycling, while also reducing waste and chemical pollution. Encourage sustainable business models that offer products as services, share and lease plastic products, and promote reuse. Recycling end-of-life plastics into new products can greatly lessen their environmental impact (Forum of the Future, 2024).

(ii) **Utilize Plastic Waste as a Resource**.

Plastic garbage can be recovered and remanufactured into various items, including bricks, composites, road construction, furniture, and clothing. Plastic trash can be transformed into liquid fuel and burned in a waste-to-energy cycle, but there are certain drawbacks. Chemical recycling allows for the recovery of petrochemical components from plastic polymers, which can be used to make new plastics, chemicals, or alternative fuels (Panda et al., 2010; Wong et al., 2015).

(iii) **Improve the Lifetime, Reusability, and Waste Prevention of Plastics by Including After-Use, Asset Recovery, and Waste and Pollution Prevention into the Design from the Beginning**.

Adopting a life-cycle approach involves cleaner production; reducing single-use plastics; designing products for extended use and easy separation, repair, upgrade, and recycling; eliminating toxic substances; and preventing microplastics from entering the environment through product redesigns. Examples include reducing wear and tear on clothing and tires, as well as removing or replacing microplastics in personal care items like toothpaste and shampoo. One example of the redesign is the bulk distribution of cleaning and personal care items in refillable plastic containers, which eliminates single-use bottles. This model has already been used for replenishing bottles and packing PET containers. A returnable bottle system and refillable bottles can save material costs and waste and reduce greenhouse gas emissions (den Hollander et al., 2017; Kaur et al., 2018).

(iv) **Promote Sustainable Business Models that View Products as Services and Encourage the Sharing and Leasing of Plastic Items**.

This approach would optimize product use, enhance revenue, and reduce manufacturing volume. For instance, households and offices can lease water dispensers and refillable plastic bottles. Lego's play concept allows buyers to rent and return sets instead of purchasing them (Forum of the Future, 2024).

(v) **Policy Instruments, such as Fiscal and Regulatory Measures, Can Address the Negative Impacts of Unsustainable Plastic Production and Consumption**.

Without these controls, markets may still prefer fossil feedstocks, especially at low oil prices, making establishing the circular economy more challenging. Considering the costs of unsustainable production and usage can drive production from less damaging sources, reduce waste, and promote reuse and recycling. Fiscal policy measures, such as direct surcharges, levies, carbon or resource taxes, and taxes on specific types of plastic (e.g., plastic bags, disposable cutlery), can discourage non-essential plastic use and improve the financial viability and quality of plastic recycling. Additional regulatory and policy measures, such as recycling targets, producer responsibility, container deposit legislation, circular/eco-design standards, public procurement policies, landfill and incineration bans, and outright bans on certain plastic products, such as single-use plastic bags, are necessary (Brink et al., 2017).

6 Conclusions

Plastics are evolving into "smart" materials, poised to play crucial roles in future living. Potential applications include human tissue or organ transplants, lightweight materials for ultra-low-emission vehicles and aircraft, advanced insulation for photovoltaic-powered homes, reusable electronic media for books and magazines, smart packaging to monitor food spoilage, and high-efficiency solid-state lighting using plastic organic diode technology. Recycling plastic waste and prolonging the lifespan of plastics support the attainment of various Sustainable Development Goals (SDGs), including those related to climate change, responsible production and consumption, and life below the ocean. Plastics are highly efficient materials that typically help reduce greenhouse gas emissions compared to common alternatives, and recycling and reusing plastics further amplify these reductions. Both mechanical and chemical recycling systems decrease greenhouse gas emissions when compared to conventional production. Recycling lessens the need for manufacturing, transportation, and raw material extraction involved in plastic production, along with the associated emissions, thereby contributing to the SDGs on climate change through these emission reductions.

References

Ahmed, N. (2023). Utilizing plastic waste in the building and construction industry. A pathway towards the circular economy. *Construction and Building Materials, 383*, 131311. https://doi.org/10.1016/j.conbuildmat.2023.131311.

Amasuomo, E., & Baird, J. (2016). The concept of waste and waste management. *Journal of Management and Sustainability, 6*(4), 88. https://doi.org/10.5539/jms.v6n4p88.

Andrady, A.L., Neal, M.A. (2009). Applications and societal benefits of plastics. *Philosophical Transactions of the Royal Society B. Biological Sciences, 364*(1526), 1977–1984. https://doi.org/10.1098/rstb.2008.0304.

Anwar, M.K., Shah, S.A.R., Alhazmi, H. (2021). Recycling and utilization of polymers for road construction projects. An application of the circular economy concept. *Polymers, 13*(8), 1330. https://doi.org/10.3390/polym13081330.

Babaremu, K. O., Okoya, S. A., Hughes, E., Tijani, B., Teidi, D., Akpan, A., Igwe, J., Karera, S., Oyinlola, M., & Akinlabi, E. T. (2022). Sustainable plastic waste management in a circular economy. *Heliyon, 8*(7), Article e09984. https://doi.org/10.1016/j.heliyon.2022.e09984.

Carbonel, D., Garriazo, Y., Mayhua, M., Orozco, S., & Tejaswini , M. S. S. R. (2023). Environmental damages due to mismanagement of municipal solid waste. anthropogenic environmental hazards. *Compensation and Mitigation,* 161–182. https://doi.org/10.1007/978-3-031-41013-0_8.

Circular Economy for Plastics in India: A Roadmap. (2021). Retrieved December 21, 2024, from www.teriin.org.

Corraini, N. R., de Souza de Lima, A., Bonetti, J., & Rangel-Buitrago, N. (2018). Troubles in the paradise: Litter and its scenic impact on the North Santa Catarina island beaches, Brazil. *Marine Pollution Bulletin, 131*, 572–579. https://doi.org/10.1016/j.marpolbul.2018.04.061.

den Hollander, M. C., Bakker, C. A., & Hultink, E. J. (2017). Product design in a circular economy-development of a typology of key concepts and terms. *Journal of Industrial Ecology, 21*(3), 517–525. https://doi.org/10.1111/jiec.12610.

Plastics Europe, Plastics–the Facts. (2022). Retrieved December 24, 2024, from https://plasticseurope.org/knowledge-hub/plastics-the-facts-2022.

Evode, N., Qamar, S. A., Bilal, M., Barceló, D., & Iqbal, H. M. N. (2021). Plastic waste and its management strategies for environmental sustainability. *Case Studies in Chemical and Environmental Engineering, 4*, Article 100142. https://doi.org/10.1016/j.cscee.2021.100142.

Forum for the future .(2024). Retrieved December 21, 2024, from https://www.forumforthefuture.org/sites/default/files/Card%20deck.pdf.

Gupta, P. (2021). Plastic waste management- a concern for community. *Holistic Approach to Environment, 11*(2), 49–66. https://doi.org/10.33765/thate.11.2.3.

Haram, L. E., Carlton, J. T., Ruiz, G. M., & Maximenko, N. A. (2020). A plasticene lexicon. *Marine Pollution Bulletin, 150*, 110714. https://doi.org/10.1016/j.marpolbul.2019.110714.

Hassanpour, M., Unnisa, S. A. (2017). Plastics; Applications, materials, processing and techniques. *Mod Tech, 2017*(02), 109. https://doi.org/10.13140/RG.2.2.15142.24647.

Ilyas, M., Ahmad, W., Khan, H., Yousaf, S., Khan, K., & Nazir, S. (2018). Plastic waste as a significant threat to environment-A systematic literature review. *Reviews on Environmental Health, 33*(4), 383–406. https://doi.org/10.1515/reveh-2017-0035.

Jones, K. H., Pathak, P. (2024). Recycling technologies and resource recovery from plastic waste. *Valorization of Resources from Urban Mined Materials. World Scientific*, 61–80. https://doi.org/10.1142/9789811297755_0004.

Karmakar, G. P. (2022). Regeneration and recovery of plastics. *Encyclopedia of Materials Plastics and Polymers, 3*, 634–651. https://doi.org/10.1016/B978-0-12-820352-1.00045-6.

Kaur, G., Uisan, K., Ong, K. L., & Ki Lin, C. S. (2018). Recent trends in green and sustainable chemistry and waste valorisation: Rethinking plastics in a circular economy. *Current Opinion in Green and Sustainable Chemistry, 9*, 30–39. https://doi.org/10.1016/j.cogsc.2017.11.003.

Kehinde, O., Ramonu, O. J., Babaremu, K. O., & Justin, L. D. (2020). Plastic wastes-environmental hazard and instrument for wealth creation in Nigeria. *Heliyon, 6*(10), Article e05131. https://doi.org/10.1016/j.heliyon.2020.e05131.

Kibria, M. G., Masuk, N. I., Safayet, R., Nguyen, H. Q., & Mourshed M. (2023). Plastic waste: Challenges and opportunities to mitigate pollution and effective management. *International Journal of Environmental Research, 17*(1). https://doi.org/10.1007/s41742-023-00507-z.

Millican, J. M., & Agarwal, S. (2021). Plastic pollution: A material problem? *Macromolecules. American Chemical Society, 54*(10), 4455–4469. https://doi.org/10.1021/acs.macromol.0c02814.

Nandikes, G., Banerjee, O., Mirthipati, M., Bhargavi, A., Jones, H., & Pathak, P. (2024). Separation, identification, and quantification of microplastics in environmental samples. *American Chemical Society, 1482*, 1–19. https://doi.org/10.1021/bk-2024-1482.ch001.

Nayanathara Thathsarani Pilapitiya, P. G. C., & Ratnayake, A. S. (2024). The world of plastic waste: A review. *Cleaner Materials, 11*, 100220. https://doi.org/10.1016/j.clema.2024.100220.

Pan, D., Su, F., Liu, C., & Guo, Z. (2021). Research progress for plastic waste management and manufacture of value-added products. *Taylor and Francis, 26*(14), 6941–6974. https://doi.org/10.1007/s42114-020-00190-0/Published.

Panda, A. K., Singh, R. K., & Mishra, D. K. (2010). Thermolysis of waste plastics to liquid fuel. A suitable method for plastic waste management and manufacture of value-added products-A world prospective. *Renewable and Sustainable Energy Reviews, 14*(1), 233–248. https://doi.org/10.1016/j.rser.2009.07.005.

Mismanaged Plastic waste per capita, Our world in data. (2019). Retrieved January, 6, 2025 from https://ourworldindata.org/.

Plastics and the circular economy A STAP document. (2018). Retrieved December 24, 2024, from http://www.stapgef.org.

Pujara, Y., Govani, J., Patel, H. T., Pathak, P., Mashru, D., & Ganesh, P. S. (2023). Quantification of environmental impacts associated with municipal solid waste management in Rajkot city, India using life cycle assessment. *Environmental Advances, 12*, Article 100364. https://doi.org/10.1016/j.envadv.2023.100364.

Shah, H. H., Amin, M., Iqbal, A., Nadeem, I., Kalin, M., Soomar, A. M., & Galal, A. M. (2023). A review on gasification and pyrolysis of waste plastics. *Frontiers in Chemistry, 10*. https://doi.org/10.3389/fchem.2022.960894.

Shanker, R., Khan, D., Hossain, R., Islam, M. T, Locock, K., Ghose, A., Sahajwalla, V., Schandl, H., & Dhodapkar, R. (2023). Plastic waste recycling: existing Indian scenario and future opportunities. *International Journal of Environmental Science and Technology, 20*(5), 5895–5912. Institute for Ionics. https://doi.org/10.1007/s13762-022-04079-x.

Tejaswini, M. S. S. R., Pathak, P., Ramkrishna, S., & Ganesh, P. S. (2022). A comprehensive review on integrative approach for sustainable management of plastic waste and its associated externalities. *Science of the Total Environment, 825*, Article 153973. https://doi.org/10.1016/j.scitotenv.2022.153973.

Brink Ten, P., Watkins, E., De Smet, M. (2017). *Circular economy measures to keep plastics and their value in the economy, avoid waste and reduce marine litter*. www.G20-insights.org.

Torres Agullo, A., Karanasiou, A., Moreno, T., & Lacorte, S. (2021). Overview on the occurrence of microplastics in air and implications from the use of face masks during the COVID-19 pandemic. *Science of the Total Environment, 800*, Article 149555. https://doi.org/10.1016/j.scitotenv.2021.149555.

Verma, R., Vinoda, K. S., Papireddy, M., & Gowda, A. N. S. (2016). Toxic pollutants from plastic waste-a review. *Procedia Environmental Sciences, 35*, 701–708. https://doi.org/10.1016/j.proenv.2016.07.069.

Villarrubia-Gómez, P., Cornell, S. E., & Fabres, J. (2018). Marine plastic pollution as a planetary boundary threat–The drifting piece in the sustainability puzzle. *Marine Policy, 96*, 213–220. https://doi.org/10.1016/j.marpol.2017.11.035.

Wong, S. L., Ngadi, N., Abdullah, T. A. T., & Inuwa, I. M. (2015). Current state and future prospects of plastic waste as source of fuel. A review. *Renewable and Sustainable Energy Reviews, 50*, 1167–1180. https://doi.org/10.1016/j.rser.2015.04.063.

Zhang, Y., Kang, S., Allen, S., Allen, D., Gao, T., & Sillanpää, M. (2020). Atmospheric microplastics: A review on current status and perspectives. *Earth-Science Reviews, 203*, Article 103118. https://doi.org/10.1016/j.earscirev.2020.103118.

Micro- and Nano-Plastics (MNPs) in the Environment: Addressing the Sampling and Identification Issues

Sadia Ilyas and Muhammad Farhan

Abstract Plastics are extensively used across various sectors, including agriculture, industries, construction, and daily life, owing to their low cost and excellent corrosion resistance. However, this widespread use has led to a significant accumulation of plastics released into the environment, coupled with growing concerns about its potential harm to biota has made plastic waste an escalating issue worldwide. Under environmental stress, plastics break down into micro- and nano-plastics (MNPs), which vary in size, shape, morphology, and composition. Despite growing concerns, research on effective methods for collecting and separating MNPs remains limited. This chapter outlines the sampling procedures for MNPs from both environmental and biological sources. It then explores the methods for extracting and separating MNPs, followed by an in-depth examination of the techniques used to identify and quantify these particles. Finally, the review discusses the need for standardized and harmonized approaches for the separation, extraction, and identification of MNPs, aiming to provide a framework for advancing pollution monitoring and risk assessment of MNPs in the future.

Keywords Micro-plastics · Nano-plastics · Sampling methodology · Separation process

1 Introduction

Plastics have largely replaced traditional materials in modern society, becoming an indispensable part of our daily lives. Under a broad term of "Plastic", the International Union of Pure and Applied Chemistry (IUPAC) defines it as "the polymeric materials that may contain other substances to enhance performance and/or reduce costs"

S. Ilyas (✉) · M. Farhan
Process Metallurgy, Department of Civil, Environmental and Natural Resources Engineering, Luleå University of Technology, Luleå, Sweden
e-mail: sadia.ilyas@ltu.se

Wallenberg Initiative Materials Science for Sustainability (WISE), Luleå University of Technology, Luleå, Sweden

P. Pathak et al. (eds.), *Plastic Footprint*, https://doi.org/10.1007/978-981-96-6362-0_2

(Vert et al., 2012). Their affordability, versatility, durability, and lightweight properties have driven a dramatic increase in global production to meet the ever-growing demand for these materials. Over the past few decades, global plastic production has surged, reaching 400 million tons in 2022, driven by the increasing demand from manufacturing industries (Plastics–the Fast Facts, 2023). Given their ubiquity in modern society, it is no surprise that some researchers refer to the present era as "the Plastic Age" (Kim & Ilyas, 2022). As a result of a well-developed economy and increasing consumer demand, a substantial amount of post-consumer plastic waste is generated each year (Chen et al., 2021). Some of these plastics exist at the nanoscale and are difficult to detect, while others break down into tiny particles due to external factors once they enter aquatic environments. These small plastic particles possess a large surface area, which allows them to adsorb a wide range of pollutants. In addition to organic contaminants, they also capture inorganic pollutants such as heavy metals and microbial contaminants in the form of biofilms (Xu et al., 2023).

In 2004, the concept of "microplastics" (MPs) was proposed (Thompson et al., 2004), defining them as particles <5 mm in diameter, based on their analysis of plastic debris in marine waters and sediments. However, MPs are not the final stage of plastic degradation; they can further break down into nano-plastics (NPs) (Sussarellu et al., 2016), of particles ranging in the size from 1 to 1000 nm (Galgani & Loiselle, 2021). The growing concerns over micro- and nano-plastics (MNPs) stem from their potential harmful effects on the growth and reproduction of plants and animals, as well as the risks they pose to human health (Chen et al., 2022; Liang et al., 2024). Plastic pollution can cause both physical and chemical harm to biota, either directly or through the sorption, concentration, and transfer of other environmental contaminants that are ingested along with the plastic (Nguyen et al., 2019). The schematic pathways through which MNPs enter the environment, as illustrated by Luo et al. (2025), are shown in Fig. 1. These pathways include industrial production, transportation, wastewater treatment plant discharges, and agricultural activities. MNPs are widely distributed in water, soil, and air, resulting in both direct and indirect ingestions by organisms during their life activities. Their small particle size is particularly concerning, as it provides a high specific surface area for the sorption of contaminants (Nguyen et al., 2019). Additionally, their small size increases their potential to translocate within the bodies of organisms, amplifying the risks they pose. Researchers have conducted extensive studies on MNP's impact on water-bodies worldwide, revealing significant variations in pollution levels across different water bodies (Shi et al., 2024; Jiang et al., 2022; Materic et al., 2022), and soils (Yao & Zhou, 2024; Zhang et al., 2022), and biota (Ilyas et al., 2022). Despite the growing evidence of MNPs across various environmental media in recent years, our understanding of their widespread presence remains limited, making the separation and detection of these particles a challenging task. Furthermore, research on the detection of nano-plastics (NPs) remains limited, with insufficient exploration of methods for extracting, separating, identifying, and quantifying NPs. Henceforth, in this chapter, we aimed to address the issues for the fulfillment of the research gaps and providing a comprehensive guide for the collection, separation, extraction, and detection of MNPs in both environmental and biological compartments.

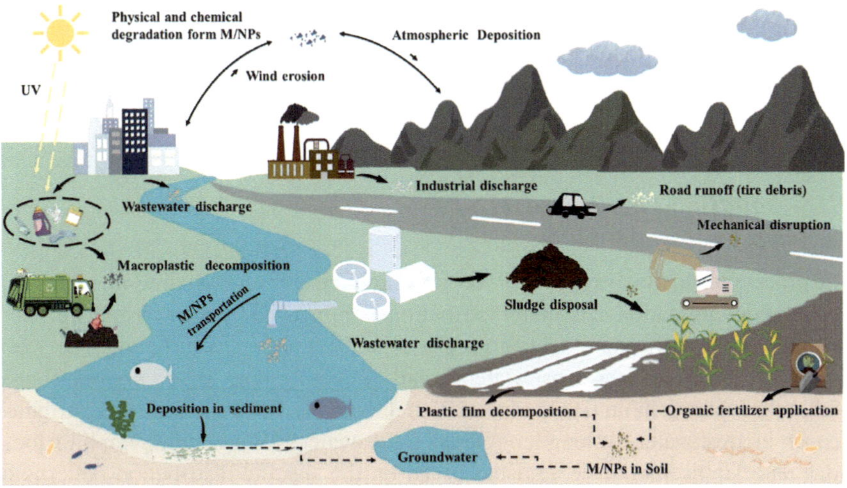

Fig. 1 Release and cycling of MNPs among environmental and biological compartments (adopted from Luo et al., 2025, with copyright permission from Elsevier)

2 Sampling of MNPs

In general, the three main sampling methodologies used for MNPs are: (i) selective sampling, (ii) volume-reduced sampling, and (iii) bulk sampling, as discussed below. Each of these methods offers specific advantages and limitations, depending on the sampling environment and the characteristics of the microplastic or nano-plastic particles being studied (Kim et al., 2022).

Selective Sampling: This method involves directly extracting visible plastic items from the environment, typically from the surface of sediments. It is commonly used for collecting plastic pellets, which are easily identifiable due to their size (1–6 mm in diameter) and presence in flotsam deposits on sandy beaches. However, selective sampling carries the risk of missing microplastics, especially when they are mixed with other debris or are irregularly shaped (e.g., rough, angular fragments). Therefore, extra care is needed when selectively sampling in the field.

Bulk Samples: This method involves collecting the entire sample volume without reducing it during the sampling process. Bulk sampling is most effective when microplastics are difficult to identify visually due to factors such as overlapping sediment particles that obscure the plastics; low abundance, requiring the sorting or filtering of large volumes of sediment or water; and small size, making them impossible to identify with the naked eye.

Volume-Reduced Samples: In this method, the volume of the bulk sample is reduced during the sampling process, retaining only the portion of interest for further analysis. For sedimentary environments, samples can be sieved directly on the beach or onboard a vessel. In seawater environments, volume-reduced samples are typically collected by filtering large volumes of water through nets. Both bulk and volume-reduced samples require additional processing in the laboratory.

2.1 Aquatic Sampling of MNPs

The procedures for analyzing MNPs broadly include sampling, separation, identification, and quantification, and their distribution is influenced by their physical properties and various environmental factors. For example, the density differences between freshwater, seawater, and wastewater, greatly affecting how MNPs are distributed within the water column (Prata et al., 2019). Generally, MNPs tend to be distributed deeper in freshwater systems compared to seawater. As a result, sampling locations and depths should be adjusted based on the specific distribution patterns of MNPs. Compared to wet environmental matrices, plastics are generally less dense and more hydrophobic (Cole & Galloway, 2015; Geyer et al., 2017). Additionally, when matrices have low solid content, plastics can be effectively separated using size-based filtration methods.

Primarily using the net-sampling in two-third cases for large water volumes, the nets of mesh size 20–800 μm and 1.5–4.5 m in length are being used to collect samples from both surface and bottom layers (Kang et al., 2015; Lima et al., 2014). Various types of nets are employed for this purpose, including plankton nets, manta trawls, catamarans, neuston nets, manual nets, and continuous nets. Plankton nets and trawl nets are typically towed by vessels, with flowmeters installed on the trawl to measure the average flow rate during the sampling period, thereby quantifying the volume of water passing through the net (Xiong et al., 2019). Plankton nets are generally towed at low speeds for short-term surface water sampling, due to their smaller mesh size and susceptibility to clogging (Prata et al., 2019). In contrast, trawl nets, with larger mesh sizes, are typically used for long-term oceanic sampling; they are incapable of collecting (NPs) from environmental waters, as reducing the mesh size <1 μm is extremely challenging (Prata et al., 2019). On the other hand, the use of bulk sampling employs glass or steel buckets to collect samples without specialized equipment or vessels, minimizing the risk of contamination. This method is particularly suitable for collecting submicron plastics and NPs, as it captures MNPs of all sizes (Luo et al., 2025; Prata et al., 2019), albeit this type of sampling has limitations, such as smaller sampling volumes and being more time-consuming. To overcome these drawbacks, some studies combine net and bulk sampling methods to leverage the advantages of both approaches (Choong et al., 2021; Hossain et al., 2022).

2.2 Sediment/Soil Sampling of MNPs

The distribution of micro- and nano-plastics (MNPs) in sediments and soil is uneven, primarily influenced by their inherent properties and environmental factors such as wind and currents. As a result, it is crucial to define the sampling area, location, depth, and other parameters. Studies examining microplastic (MP) densities in sedimentary environments often lack consistency in assessing MP accumulation across different beach zones. In many of these studies, selective sediment samples for plastic pellets and fragments have been collected using tools such as tweezers (Ashton et al., 2010), tablespoons (Cooper & Corcoran, 2010), or even by hand (Mato et al., 2001; Turner & Holmes, 2011). Some studies focused on the high tide line, employing various sampling methods, including the sampling of a linear extension along the strandline with a spoon and/or trowel, using quadrats to sample an aerial extension, and collecting samples from different depth strata using corers.

Random sampling is used to identify contamination in a site assumed to be uniformly contaminated, while systematic grid sampling helps assess the extent of contamination, delineate hotspots, and identify spatial concentration gradients. Systematic random sampling is employed to determine the overall pollution level across the survey area (Möller et al., 2020). The sub-sampling points are then established using methods such as random sampling (Xu et al., 2022), plum blossom method, or snake method. Samples collected from these sub-sampling points are mixed in equal amounts to ensure representativeness. In heavily polluted areas, stratified sampling may be used to capture a more accurate depiction of the contamination. For soil sampling, the typical collection depth is 0–20 cm for surface soil, 20–30 cm for subsurface soil, and deeper than 30 cm for deep soil samples (Chia et al., 2024). The typical sample size is 1–2 kg, with stratified samples usually around 100–200 g. To avoid contamination and ensure the accuracy of the results, sample collection tools are typically made of metal. Stainless steel shovels and spoons are commonly used for surface sampling, while earth drills are preferred for deeper soil sampling. Samples are typically stored in aluminum foil bags or boxes to preserve their integrity.

2.3 Atmospheric Sampling of MNPs

The atmospheric sampling of MNPs can be categorized into passive and active sampling, depending on the objectives and purposes of the study. Passive sampling relies on the principle of molecular diffusion to enable long-term collection of MNPs from the atmosphere. While techniques such as sweeping, vacuuming, and active pumping are relatively easy to perform, comparing data across different studies is challenging, particularly when determining the relative quantity of sampled air or confirming whether the collected microplastics are exclusively from atmospheric deposition (Liu et al., 2019). Recently, the Norwegian Institute for Air Research has developed a standardized passive atmospheric deposition of microplastics. This method, which is plastic-free and easy to use, requires no power and collects a known volume of blow-by samples. This standardization allows for reliable comparisons to other deposited materials and facilitates comparisons across various plastic studies. Active air sampling, which involves pumping air through a filter, is another established protocol tailored to specific meteorological conditions and surface environments, whether terrestrial or oceanic. It is a widely used method for atmospheric pollution monitoring (Dommergue et al., 2019; Hayward et al., 2010). Active sampling efficiently collects atmospheric MNPs from both outdoor and indoor air, but its effectiveness can be significantly impacted by weather conditions (Luo et al., 2025).

3 Separation of MNPs

Like other physical separation processes, isolation of plastics from complex matrices also applies the distinctive properties of plastics that set them apart from their surrounding environment of the bulk either solid or liquid. Figure 2 depicts the application of density and size-based separation methods from different samples, including the density separation, dead-end filtration, and froth filtration techniques. The density separation process leverages the properties of plastic materials to separate substances based on their differing densities (refer to Table 1 summarizing the density of common plastics), such as sediments and MPs. When a slurry of intermediate density is used, materials with lower densities will float to the top, while those with higher densities will sink to the bottom. The floating MPs can then be collected by decanting the upper layer, leaving the sediments at the bottom. To alter the density of MPs from the bulk, a salt solution like brine is used that allows plastic to float, facilitating a faster separation. Additionally, by adjusting the ethanol-to-water ratio, solutions with densities lower than that of water can be prepared.

On the other side, the froth filtration process relies on the selective separation of materials based on their hydrophobic or hydrophilic properties, based on the surface wettability. The materials with hydrophobic surfaces tend to aggregate with air bubbles and float, while hydrophilic materials remain in the liquid phase and are carried away as underflow. Plastic waste typically has hydrophobic, low-energy surfaces, allowing them to adhere to bubbles and be lifted to the surface, forming a froth that can be removed. In contrast, hydrophilic materials stay in the liquid phase. Since plastics are inherently hydrophobic, they can be effectively separated using froth flotation. Furthermore, to separate two different plastic materials, a wetting agent can be added, which selectively adsorbs onto the surface of one type of plastic. This agent acts as a flotation depressant, promoting selective adsorption and altering the material's hydrophilic properties. As a result, the hydrophobic plastic will float, while the hydrophilic plastic will sink due to the reduced flotation, enabling efficient extraction of the plastics from the surface.

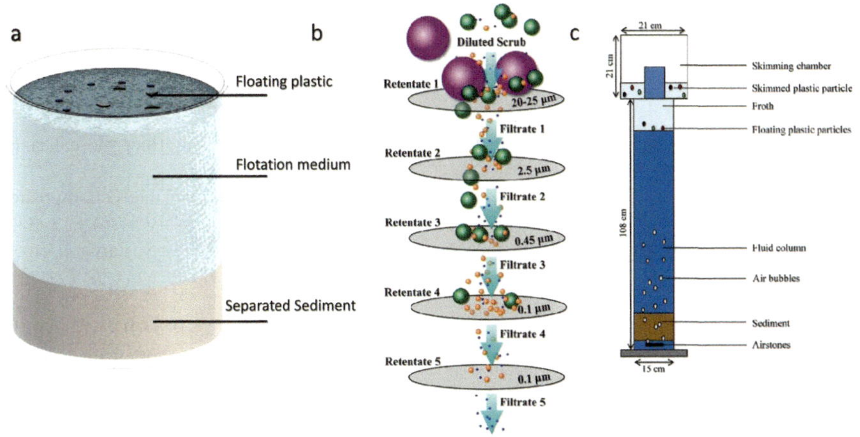

Fig. 2 The plastic separation techniques using **a** density separation, **b** dead-end filtration, and **c** froth filtration (adopted from Nguyen et al., 2019, with copyright permission from American Chemical Society)

3.1 Organisms Sampling of MNPs

As evident that MNPs in water, soil, and the atmosphere can enter to fish and human body through the food chain (Leslie et al., 2022). Hence, to understand MNPs accumulation in living organisms, appropriate sampling methods are imperative. Due to the diversity of organisms, different techniques are used. For example, to sample the aquatic organisms, ranging from phytoplankton to fish, trawls of various sizes are commonly used (Frias et al., 2014). Benthic organisms can be effectively captured using bottom trawling or sediment collection methods, while plants are typically collected manually; however, it often necessitates dissection or digestion for proper analysis (Alomar et al., 2020; Lusher et al., 2017). Acid or base digestion is commonly used to isolate MPs from biological tissues, as the protocols defined in Table 2 (Nguyen et al., 2019). Enzymatic digestion, using enzymes such as proteinase K, chitinase, or cellulase, has also been successfully applied for more targeted digestion of tissue, either alone or in combination with chemical digestants. Dissection is a traditional and cost-effective method for visual identification of MPs larger than 500 μm uptake in the gastrointestinal tracts of larger organisms (e.g., fish) or whole bodies of smaller organisms (e.g., mussel), whereas smaller MPs may translocate to other tissues and organs (Mattsson et al., 2017; Reineke et al., 2013). Fixation and cryosectioning techniques can be useful for investigating the MP's movement to other tissues, such as the liver, before further examination and quantification. It needs to highlight that to prevent contamination, the sampling tools/containers must thoroughly be rinsed with 35% ethanol solution and the collected samples should be stored and transported at −20 °C.

Table 1 List of some common plastics with density ranges and common high-density solutions to separate sediment microplastics

Plastics/abbreviation	Density (g/cm^3)
Poly (methyl methacrylate)/PMA	1.17–1.20
Poly (methyl methacrylate) (high heat)/PMA	1.15–1.25
Poly (methyl methacrylate) (impact modified)/PMA	1.10–1.20
Polytetrafluoroethylene/PTFE	2.10–2.20
Polytetrafluoroethylene (25% glass fiber)/PTFE	2.20–2.30
Polypropylene (impact modified)/PP	0.88–0.91
Polypropylene (homopolymer)/PP	0.90–0.91
Polypropylene (copolymer)/PP	0.90–0.91
Polypropylene (10–20% glass fiber)/PP	0.97–1.05
Polypropylene (10–40% mineral filled)/PP	0.97–1.25
Polypropylene (10–40% talc)/PP	0.97–1.25
Polypropylene (30–40% glass fiber)/PP	1.10–1.23
Low-density polyethylene/LDPE	0.92–0.94
Linear low-density polyethylene/LLDPE	0.92–0.95
High-density polyethylene/HDPE	0.94–0.97
Polystyrene (crystal)/PS	1.04–1.05
Polystyrene (high heat)/PS	1.04–1.05
Acrylonitrile butadiene styrene/ABS	1.03–1.21
Acrylonitrile butadiene styrene (flame retardant)/ABS	1.15–1.20
Acrylonitrile butadiene styrene (high impact)/ABS	1.00–1.10
Acrylonitrile butadiene styrene (high heat)/ABS	1.00–1.15
Nylon 6,6 (impact modified)/PA	1.05–1.10
Nylon 6/PA	1.12–1.14
Nylon 6,6/PA	1.13–1.15
Nylon 6,6 (30% glass fiber)/PA	1.37–1.38
Nylon 6,6 (impact modified and 15–30% glass fiber)/PA	1.25–1.35
Nylon 6,6 (30% mineral filled)/PA	1.35–1.38
Polystyrene (expanded foam)/EPS	0.01–0.05
Polystyrene (extruded foam)/XPS	0.03–0.05
Polychloroprene (neoprene) (solid)/CR	1.20–1.24
Polychloroprene (neoprene) (foamed)/CR	0.11–0.56
Polyethylene terephthalate/PET	1.30–1.40
Polyethylene terephthalate (30% glass fiber)/PET	1.50–1.60
Polyethylene terephthalate (30% glass fiber and impact modified)/PET	1.40–1.50
Polycarbonate (high heat)/PC	1.15–1.20
Polycarbonate (20–40% glass fiber and flame retardant)/PC	1.40–1.50

(continued)

Table 1 (continued)

Plastics/abbreviation	Density (g/cm^3)
Polycarbonate (20–40% glass fiber)/PC	1.35–1.52
Polystyrene (30% glass fiber)/PS	1.40–1.50
Polyvinyl chloride (rigid)/PVC	1.35–1.50
Polyvinyl chloride (20% glass fiber)/PVC	1.45–1.50
Polyvinyl chloride (plasticized)/PVC	1.30–1.70
Polyvinyl chloride (plasticized and filled)/PVC	1.15–1.35
Salt solution/chemical formula	*Density* (g/cm^3)
Sodium iodide/NaI	1.8
Zinc chloride/$ZnCl_2$	1.5–1.7
Sodium polytungstate/Na_6(H2W12O40)	1.4
Lithium metatungstate (5.4 mol/L)/$Li_6(H_2W_{12}O_{40})$	1.6

4 Detection of MNPs

Numerous advanced techniques have been developed to identify MNPs in samples from various environments, as discussed in earlier sections. The choice of characterization method depends on the particle size of the MNPs and the nature of the sample. These techniques include visual detection using high-resolution electron microscopy, vibrational spectroscopy, and various mass spectrometric analyses. Figure 3 illustrates the classification of different methodologies employed for detecting MNPs in various sample types (Luo et al., 2025).

4.1 Visual Detection Using High-Resolution Electron Microscopes

The visual characterization techniques provide a cost-effective and efficient means of reducing the number of particles that need subsequent chemical analysis (Renner et al., 2018). This approach involves examining physical attributes of particles, such as morphology and color, that are characteristic of plastic. For particles larger than approximately 500 μm, visual identification using the naked eye or light microscopy is typically sufficient (Rocha-Santos et al., 2015). Conventional light microscopy, even with optimal optics, is limited by the fundamental diffraction limit of approximately 200 nm, restricting its ability to resolve smaller particles. For instance, a scanning electron microscope (SEM) generates high-resolution images of a sample's surface by directing a high-intensity electron beam onto it, which scans in a zig-zag raster pattern. Because SEM primarily focuses on surface imaging, it can achieve a resolution as fine as 0.5 nm at magnifications up to $\times$ 2,000,000. This level of magnification far exceeds that of a standard optical microscope, which is typically

Table 2 List of different protocols for biological sample digestion in wastewater microplastic studies (adopted from Nguyen et al., 2019, with copyright permission from American Chemical Society)

Sample type	Processing steps					
Influent	Sieving					
Pretreatment effluent	Sieving					
Secondary effluent	Sieving					
Tertiary effluent	Sieving					
Anaerobic digestor sludge	Sieving					
Dewatering reject water	Sieving					
Dried sludge	Sieving					
Tertiary effluent	Sieving					
Secondary effluent	Sieving					
Pretreatment effluent	Sieving					
Grit and grease	Sieving					
Primary effluent	Sieving					
Secondary effluent	Sieving					
Anaerobic biological process	Density separation	H_2O_2 or Fenton digestion				
Secondary and tertiary effluents	Sieving	Density separation				
Bio-solids	Density separation	Bleach oxidation				
Primary effluent	Sieving	Density separation	Filtrate			
		H_2O_2 or Fenton digestion	Retentate			
Secondary effluent	Sieving	Density separation	Filtrate			

(continued)

Table 2 (continued)

Sample type	Processing steps					
		H_2O_2 or Fenton digestion	Retentate			
Primary sludge	Density separation	H_2O_2 or Fenton digestion	Settled sludge			
		Density separation	Supernatant			
Secondary sludge	Density separation	H_2O_2 or Fenton digestion	Settled sludge			
		Density separation	Supernatant			
Secondary effluent	Sieving	H_2O_2 or Fenton digestion	Sieving			
Raw influent	Sieving	Enzymatic digestion	H_2O_2 or Fenton digestion	Sieving		
Secondary effluent	Sieving	Enzymatic digestion	H_2O_2 or Fenton digestion	Sieving		
Dewatered sludge	Density separation	Sieving	H_2O_2 or Fenton digestion	Sieving		
Sludge, lime stabilization	Homogenization	Sieving	Sieving			
Sludge, thermal drying	Homogenization	Sieving	Density separation	Sieving		
Sludge, anaerobic digestion	Homogenization	Sieving	Density separation	sieving		
Sludge, thermal drying + Anaerobic digestion	Homogenization	Sieving	Density separation	Sieving		
Primary effluent	Sieving	Heat concentration	H_2O_2 or Fenton digestion	Density separation	Sieving	Staining

(continued)

Table 2 (continued)

Sample type	Processing steps					
Secondary effluent	Sieving	Heat concentration	H_2O_2 or Fenton digestion	Density separation	Sieving	Staining
Tertiary effluent	Sieving	Heat concentration	H_2O_2 or Fenton digestion	Density separation	Sieving	Staining
Tertiary effluent	Sieving	Heat concentration	H_2O_2 or Fenton digestion	Density separation	Sieving	Staining

limited to a maximum magnification of around × 1,000. Additionally, the X-rays are produced when the electron beam interacts with the sample surface. These X-rays are then detected by an X-ray microanalyzer, which converts their emission intensity into voltage through energy-dispersive X-ray (EDX) analysis. This technique enables the precise determination of the sample's elemental composition (Eriksen et al., 2013; Vandermeersch et al., 2015). However, it is hindered by its high cost, intricate sample preparation, and inability to definitively confirm the presence of plastic. A transmission electron microscopy (TEM) is similarly unsuitable for visualizing MNPs due to their amorphous structure, necessitating the use of heavy-metal stains. In contrast, noninvasive methods such as two-photon excitation and time-resolved detection have been employed to observe aggregated 12 nm euluminescence-labeled NPs within *Musa exotica*, offering a simpler sample preparation process compared to electron microscopy (Nguyen et al., 2019). But these methods are still constrained by diffraction limits. By leveraging the principles of Brownian motion and light scattering, NTA can detect particles as small as 30 nm.

4.2 Visual Detection Using Vibrational Spectroscopy

Although vibrational spectroscopy coupled with optical microscopy is highly effective in identifying specific types of MNPs, the signal obtained is influenced by particle size and typically requires well-separated samples (Harshvardhan & Jha, 2013). Fourier-transform infrared spectroscopy (FTIR), a non-destructive analytical technique, is commonly used to identify characteristic band patterns associated with functional groups. This technique relies on changes in the permanent dipole moment of chemical bonds, making it especially sensitive to polar functional groups, such as those found in polymers. FTIR can achieve spatial resolutions as fine as 5 μm (Elert et al., 2017). However, FTIR has certain limitations that it requires a minimum sample thickness of about 150 nm and deposition onto an IR-transparent substrate (Löder & Gerdts, 2015; Mallikarjunachari & Ghosh, 2016). Consequently, FTIR is

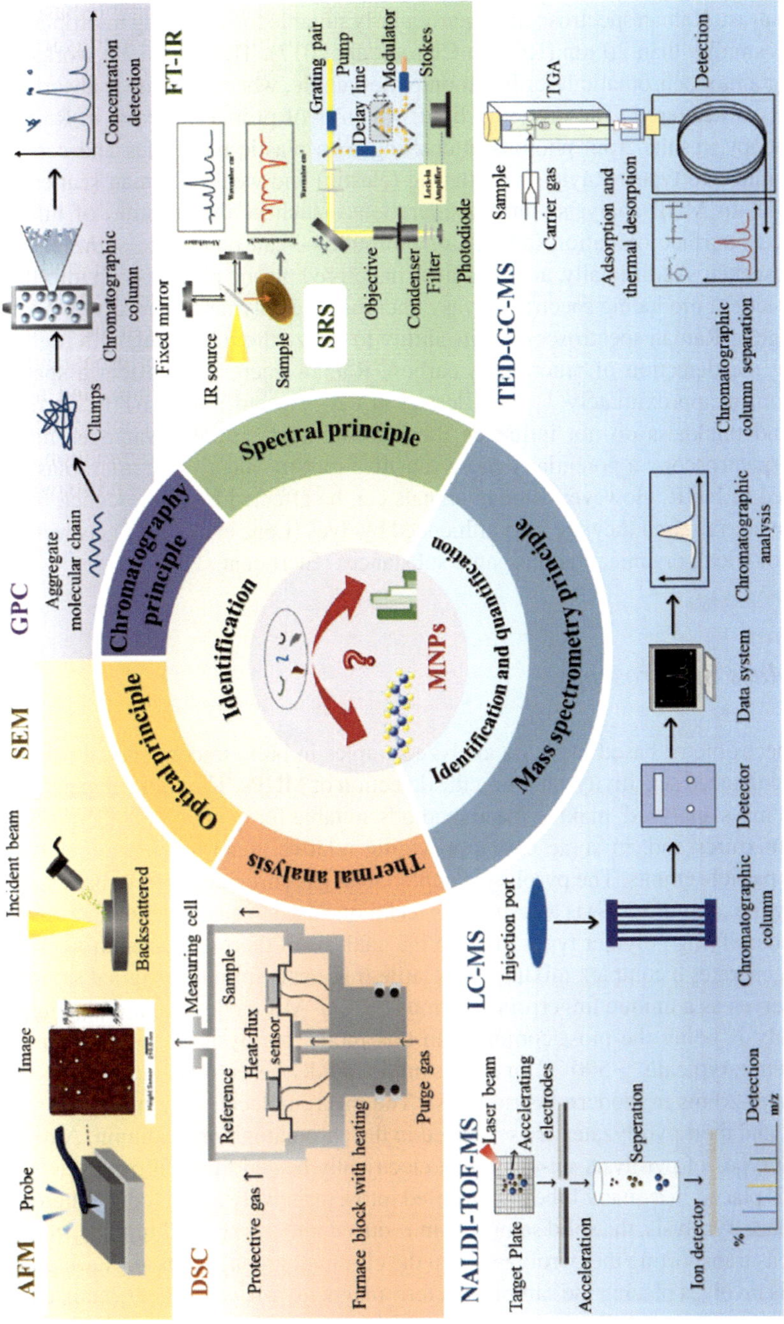

Fig. 3 Different classifications of MNP detection techniques based on analytical principles (adopted from Luo et al., 2025, with copyright permission from Elsevier)

most suitable for analyzing individual particles >20 μm, though agglomerates or films of smaller particles can also be examined.

In contrast, Raman spectroscopy is particularly suitable for analyzing microplastic particles smaller than 20 μm (Ribeiro-Claro et al., 2017). This technique works by directing a monochromatic laser beam onto the sample, where it is either absorbed, reflected, or scattered. The scattered light, which is of primary interest in Raman spectroscopy, results from photons interacting with plastic molecules and can be divided into two types: Rayleigh scattering (elastic) and weaker Raman scattering (inelastic). In MNP analysis, surface alterations—such as the sorption of humic substances, surface oxidation (aging), or biofouling—can contribute significantly to fluorescence. Additionally, artificial dyes in colored plastic particles can interfere with detection, producing spectra of dyes rather than the polymers themselves. A key advantage of Raman spectroscopy is its ability to utilize the full wavelength region, enabling the detection of amorphous carbon. Raman microscopy offers a spatial resolution of approximately 1 μm (Elert et al., 2017), and unlike FTIR, particle shape and thickness do not influence the measurement. These advantages make Raman spectroscopy a potentially more sensitive tool for identifying microplastics compared to FTIR. However, Raman signals can be affected by fluorescence from certain materials, and they are also influenced by dyes (Lenz et al., 2015), as well as microbiological, organic, and inorganic substances (Elert et al., 2017).

4.3 Mass Spectroscopy

Mass spectrometry-based methods analyze samples in bulk, trading spatial resolution for enhanced sensitivity, enabling the detection of MNPs. The signal depends on the total mass analyzed, making these methods suitable for qualitatively identifying plastic mixtures and, in some cases, providing relative quantification rather than discrete particle counts. The pyrolysis coupled with gas chromatography-mass spectroscopy (i.e., Py-GC–MS) is a destructive technique used to characterize microplastics by identifying polymer types through the analysis of thermal degradation products. It generates a complex mixture of volatile fragments from the original sample, which serves as a unique fingerprint. Various Py-GC–MS modes exist, with single-shot analysis being the most common. In this mode, pyrolysis occurs at a single temperature, typically > 500 °C, and the sample rapidly heats to the target temperature (under 20 ms in modern instruments). The macromolecules are instantly fragmented, and their pyrolyzates are separated in the chromatographic column. Among different types of pyrolysis set-ups, in an electrically heated filament pyrolysis, the sample is placed in a quartz tube and inserted into a resistively heated coil. Whereas, in a furnace pyrolysis, the solid sample is introduced into a preheated furnace, with a carrier gas transporting the pyrolyzate into the chromatography column. Curie point pyrolysis involves placing the sample in a ferromagnetic foil, which is then heated in an induction pyrolyzer. The key advantage of filament and Curie point pyrolysis is the use of disposable sample substrates, allowing suspensions of MNPs to be dried and pyrolyzed. While all of these methods result in temperature increase under a controlled atmosphere (Dworzanski & Meuzelaar, 2017).

On the other side, the thermal desorption coupled with gas chromatography-mass spectrometry (TDS-GC–MS) performs double-shot analysis, providing information on both volatile and non-volatile compounds. It involves heating a sample on a thermogravimetric balance up to 1000 °C, causing degradation products to adsorb onto a solid phase. These products are then transferred to a thermal desorption unit, desorbed by increasing the temperature, separated via chromatography, and analyzed by mass spectrometry. TDS-GC–MS is ideal for samples with relatively high mass (up to 100 mg), but it is limited to qualitative analysis (Nguyen et al., 2019).

Other than the Py-GC–MS and TDS-GC–MS, an evolved gas analysis mass spectroscopy (EGA-MS) separates degradation products based on the temperature at which they form, rather than their volatilization temperature. This is achieved by gradually heating the sample in the pyrolyzer and replacing the chromatographic column with a short, narrow deactivated capillary tube (2.5 m, 0.15 mm) that connects the GC injector directly to the MS detector. Similar to thermogravimetric analysis, EGA-MS does not measure weight loss but provides a thermogram, such as a differential thermogravimetric curve, for the MNP samples.

5 Summary

Due to their small size, low concentrations, diverse forms, and tendency to aggregate with organisms or other substances, MNP's pollution are particularly challenging to extract, separate, identify, and quantify. In this account, we have discussed recent advances in the separation and analysis of MNPs, with a particular focus on the processing of complex samples, including the aquatic and organism's sampling. Additionally, the methods for identifying and quantifying MNPs are categorized; however, a multimethod approach that combines the unique strengths of different detection techniques can be beneficial to identify the MNPs from different samples. Furthermore, the development of new methods that combine various techniques could benefit from insights gained in other fields.

Acknowledgements The authors greatly acknowledge the Wallenberg Initiative Materials Science for Sustainability (WISE) funded by the Knut and Alice Wallenberg Foundation, for supporting this work.

References

Alomar, C., Deudero, S., Compa, M., & Guijarro, B. (2020). Exploring the relation between plastic ingestion in species and its presence in seafloor bottoms. *Marine Pollution Bulletin, 160*, Article 111641.

Ashton, K., Holmes, L., & Turner, A. (2010). Association of metals with plastic production pellets in the marine environment. *Marine Pollution Bulletin, 60*(11), 2050–2055.

Chen, G., Li, Y., & Wang, J. (2021). Occurrence and ecological impact of microplastics in aquaculture ecosystems. *Chemosphere, 274*, Article 129989.

Chen, G., Li, Y., Liu, S., Junaid, M., & Wang, J. (2022). Effects of micro (nano) plastics on higher plants and the rhizosphere environment. *Science of the Total Environment, 807*, Article 150841.

Chia, R. W., Lee, J. Y., Cha, J., & Rodríguez-Seijo, A. (2024). Methods of soil sampling for microplastic analysis: A review. *Environmental Chemistry Letters, 22*(1), 227–238.

Choong, W. S., Hadibarata, T., Yuniarto, A., Tang, K. H. D., Abdullah, F., Syafrudin, M., Al Farraj, D. A., & Al-Mohaimeed, A. M. (2021). Characterization of microplastics in the water and sediment of baram river estuary, borneo island. *Marine Pollution Bulletin, 172*, Article 112880.

Cole, M., & Galloway, T. S. (2015). Ingestion of nanoplastics and microplastics by Pacific Oyster Larvae. *Environmental Science and Technology, 49*, 14625–14632.

Cooper, D. A., & Corcoran, P. L. (2010). Effects of mechanical and chemical processes on the degradation of plastic beach debris on the island of Kauai, Hawaii. *Marine Pollution Bulletin, 60*(5), 650–654.

Dommergue, A., Amato, P., Tignat-Perrier, R., Magand, O., Thollot, A., Joly, M., Bouvier, L., Sellegri, K., Vogel, T., Sonke, J. E., Jaffrezo, J. L., Andrade, M., Moreno, I., Labuschagne, C., Martin, L., Zhang, Q., & Larose, C. (2019). Methods to investigate the global atmospheric microbiome. *Frontiers in Microbiology, 10*, 243.

Dworzanski, J. P., & Meuzelaar, H. L. (2017). C. Pyrolysis mass spectrometry, methods. In Encyclopedia of Spectroscopy and Spectrometry. Elsevier, pp. 789−801.

Elert, A. M., Becker, R., Duemichen, E., Eisentraut, P., Falkenhagen, J., Sturm, H., & Braun, U. (2017). Comparison of different methods for MP detection: What can we learn from them, and why asking the right question before measurements matters? *Environmental Pollution, 231*, 1256–1264.

Eriksen, M., Maximenko, N., Thiel, M., Cummins, A., Lattin, G., Wilson, S., Hafner, J., Zellers, A., & Rifman, S. (2013). Plastic pollution in the South Pacific subtropical gyre. *Marine Pollution Bulletin, 68*(1–2), 71–76.

Frias, J. P., Otero, V., & Sobral, P. (2014). Evidence of microplastics in samples of zooplankton from Portuguese coastal waters. *Marine Environment Research, 95*, 89–95.

Galgani, L., & Loiselle, S. A. (2021). Plastic pollution impacts on marine carbon biogeochemistry. *Environmental Pollution, 268*, Article 115598.

Geyer, R., Jambeck, J. R., & Law, K. L. (2017). Production, use, and fate of all plastics ever made. *Science Advances, 3*, Article e1700782.

Harshvardhan, K., & Jha, B. (2013). Biodegradation of low-density polyethylene by marine bacteria from pelagic waters, Arabian Sea, India. *Marine Pollution Bulletin, 77*(1–2), 100–106.

Hayward, S. J., Gouin, T., & Wania, F. (2010). Comparison of four active and passive sampling techniques for pesticides in air. *Environmental Science and Technology, 44*(9), 3410–3416.

Hossain, M. J., AftabUddin, S., Akhter, F., Nusrat, N., Rahaman, A., Sikder, M. N. A., et al. (2022). Surface water, sediment, and biota: The first multi-compartment analysis of microplastics in the Karnafully river, Bangladesh. *Marine Pollution Bulletin, 180*, Article 113820.

Ilyas, S., Kim, H., & Hwang, G. (2022). Impacts of microplastics and nanoplastics on biota. In *Microplastics* (pp. 111–139). CRC Press.

Jiang, Y., Yang, F., Kazmi, S. S. U. H., Zhao, Y., Chen, M., & Wang, J. (2022). A review of microplastic pollution in seawater, sediments and organisms of the Chinese coastal and marginal seas. *Chemosphere, 286*, Article 131677.

Kang, J. H., Kwon, O. Y., Lee, K. W., Song, Y. K., & Shim, W. J. (2015). Marine neustonic microplastics around the southeastern coast of Korea. *Marine Pollution Bulletin, 96*(1–2), 304–312.

Kim, H., & Ilyas, S. (2022). Emergence, chemical nature, classification, environmental impact, and analytical challenges of various plastics. In *Microplastics* (pp. 1–30). CRC Press.

Kim, H., Ilyas, S., Hong, G., Lee, B. C., & Kim, G. (2022). Collection of microplastics and nanoplastics from various environments and associated challenges. In *Microplastics* (pp. 141–156). CRC Press.

Lenz, R., Enders, K., Stedmon, C. A., Mackenzie, D. M. A., & Nielsen, T. G. A. (2015). Critical assessment of visual identification of marine Microplastic using Raman spectroscopy for analysis improvement. *Marine Pollution Bulletin, 100*, 82–91.

Leslie, H. A., Van Velzen, M. J., Brandsma, S. H., Vethaak, A. D., Garcia-Vallejo, J. J., & Lamoree, M. H. (2022). Discovery and quantification of plastic particle pollution in human blood. *Environment International, 163*, Article 107199.

Liang, J., Ji, F., Wang, H., Zhu, T., Rubinstein, J., Worthington, R., et al. (2024). Unraveling the threat: Microplastics and nano-plastics' impact on reproductive viability across ecosystems. *Science of the Total Environment, 913*, Article 169525.

Lima, A. R. A., Costa, M. F., & Barletta, M. (2014). Distribution patterns of microplastics within the plankton of a tropical estuary. *Environmental Research, 132*, 146–155.

Liu, C., Li, J., Zhang, Y., Wang, L., Deng, J., Gao, Y., & Sun, H. (2019). Widespread distribution of PET and PC microplastics in dust in urban China and their estimated human exposure. *Environment International, 128*, 116–124.

Löder, M. G. J., & Gerdts, G. (2015). Methodology used for the detection and identification of microplastics-a critical appraisal. In *Marine Anthropogenic Litter* (pp. 201−227).

Luo, Y., Hu, R., Zheng, Y., Ding, L., Qiu, X., & Yang, J., et al. (2025). Sampling, extraction, and analysis of micro-and nano-plastics in environmental and biological compartments: A review. TrAC, Trends Analytical Chemistry, 118056.

Lusher, A. L., Welden, N. A., Sobral, P., & Cole, M. (2017). Sampling, isolating and identifying microplastics ingested by fish and invertebrates. *Analytical Methods, 9*, 1346–1360.

Mallikarjunachari, G., & Ghosh, P. (2016). Analysis of strength and response of polymer nano thin film interfaces applying nanoindentation and nanoscratch techniques. *Polymer, 90*, 53–66.

Materić, D., Peacock, M., Dean, J., Futter, M., Maximov, T., Moldan, F., et al. (2022). Presence of nanoplastics in rural and remote surface waters. *Environmental Research Letters, 17*(5), Article 054036.

Mato, Y., Isobe, T., Takada, H., Kanehiro, H., Ohtake, C., & Kaminuma, T. (2001). Plastic resin pellets as a transport medium for toxic chemicals in the marine environment. *Environmental Science and Technology, 35*(2), 318–324.

Mattsson, K., Johnson, E. V., Malmendal, A., Linse, S., Hansson, L. A., & Cedervall, T. (2017). Brain damage and behavioural disorders in fish induced by plastic nanoparticles delivered through the food chain. *Science and Reports, 7*, 11452.

Möller, J. N., Löder, M. G., & Laforsch, C. (2020). Finding microplastics in soils: A review of analytical methods. *Environmental Science and Technology, 54*(4), 2078–2090.

Nguyen, B., Claveau-Mallet, D., Hernandez, L. M., Xu, E. G., Farner, J. M., & Tufenkji, N. (2019). Separation and analysis of microplastics and nanoplastics in complex environmental samples. *Accounts of Chemical Research, 52*(4), 858–866.

Plastics–the Fast Facts. (2023). *An analysis of european plastics production, demand waste data*. https://plasticseurope.org/knowledge-hub/plastics-the-fast-facts-2023/.

Prata, J. C., Da Costa, J. P., Duarte, A. C., & Rocha-Santos, T. (2019). Methods for sampling and detection of microplastics in water and sediment: A critical review. *TrAC Trends in Analytical Chemistry, 110*, 150–159.

Reineke, J. J., Cho, D. Y., Dingle, Y. T., Morello, A. P., Jacob, J., Thanos, C. G., & Mathiowitz, E. (2013). Unique Insights into the Intestinal absorption, transit, and subsequent biodistribution of polymer-derived microspheres. *Proceedings of the National Academy of Sciences, 110*, 13803–13808.

Renner, G., Schmidt, T. C., & Schram, J. (2018). Analytical methodologies for monitoring micro(nano)plastics: Which are fit for purpose? *Current Opinion in Environmental Science & Health, 1*, 55–61.

Ribeiro-Claro, P., Nolasco, M. M., & Araújo, C. (2017). Characterization of microplastics by raman spectroscopy. *Comprehensive Analytical Chemistry, 75*, 119–151.

Rocha-Santos, T., & Duarte, A. C. A. (2015). Critical Overview of the analytical approaches to the occurrence, the fate and the behavior of microplastics in the environment. *TrAC Trends in Analytical Chemistry, 65*, 47–53.

Shi, C., Liu, Z., Yu, B., Zhang, Y., Yang, H., Han, Y., et al. (2024). Emergence of nanoplastics in the aquatic environment and possible impacts on aquatic organisms. *Science of the Total Environment, 906*, Article 167404.

Sussarellu, R., Suquet, M., Thomas, Y., Lambert, C., Fabioux, C., Pernet, M. E. J., et al. (2016). Oyster reproduction is affected by exposure to polystyrene microplastics. *Proceedings of the National Academy of Sciences, 113*(9), 2430–2435.

Thompson, R. C., Olsen, Y., Mitchell, R. P., Davis, A., Rowland, S. J., John, A. W., et al. (2004). Lost at sea: Where is all the plastic? *Science, 304*(5672), 838–838.

Turner, A., & Holmes, L. (2011). Occurrence, distribution and characteristics of beached plastic production pellets on the island of Malta (central Mediterranean). *Marine Pollution Bulletin, 62*(2), 377–381.

Vandermeersch, G., Van Cauwenberghe, L., Janssen, C. R., Marques, A., Granby, K., Fait, G., Kotterman, M. J. J., Diogène, J., Bekaert, K., Robbens, J., & Devriese, L. (2015). A critical view on microplastic quantification in aquatic organisms. *Environmental Research, 143*, 46–55.

Vert, M., Doi, Y., Hellwich, K. H., Hess, M., Hodge, P., Kubisa, P., & Schué, F. (2012). Terminology for biorelated polymers and applications (IUPAC Recommendations 2012). *Pure and Applied Chemistry, 84*(2), 377–410.

Xiong, X., Wu, C., Elser, J. J., Mei, Z., & Hao, Y. (2019). Occurrence and fate of microplastic debris in middle and lower reaches of the Yangtze River–from inland to the sea. *Science of the Total Environment, 659*, 66–73.

Xu, G., Yang, L., Xu, L., & Yang, J. (2022). Soil microplastic pollution under different land uses in tropics, southwestern China. *Chemosphere, 289*, Article 133176.

Xu, J., Wu, G., Wang, H., Ding, Z., & Xie, J. (2023). Recent study of separation and identification of micro-and nanoplastics for aquatic products. *Polymers, 15*(21), 4207.

Yao, B., & Zhou, Y. Y. (2024). Removal of neonicotinoid insecticides from water in various treatment processes: A review. *Critical Reviews in Environment Science and Technology, 54*, 1307–1339.

Zhang, Y., Zhang, X., Li, X., & He, D. (2022). Interaction of microplastics and soil animals in agricultural ecosystems. *Current Opinion in Environmental Science & Health, 26*, Article 100327.

Advancing Sustainability in Construction: Integrating Recycled Plastic Waste into Green Concrete

Muhammad Farhan, Sadia Ilyas, and Nimra Ilyas

Abstract Rapid population growth and shifts in lifestyles have significantly increased plastic production and disposal, resulting in a major environmental challenge. The persistence of plastic waste, due to its non-biodegradable nature and large volume, has become a critical concern. Recycling plastic waste into valuable products offers a more sustainable solution compared to traditional disposal methods, effectively addressing waste management issues while also enhancing economic value. One promising approach is the use of green concrete, a sustainable composite material that incorporates natural and waste-derived substances as substitutes for conventional cement and aggregates, that may be a key contributor to carbon dioxide emissions and shows the great interest in developing green concrete using readily available natural and waste materials. This chapter explores studies on cement-based materials that integrate recycled plastic waste into concrete. It also examines the potential applications of machine learning and artificial intelligence in predicting the mechanical properties of plastic-infused concrete, such as workability; reduced density; and achieving compressive, tensile, and flexural strengths within acceptable limits. It is suggested that incorporating plastic waste into construction not only promotes sustainability but also helps to mitigate environmental pollution.

Keywords Plastic waste · Recycled aggregate · Recycled fibers · Waste management · Machine learning

---

M. Farhan · S. Ilyas (✉)
Process Metallurgy, Department of Civil, Environmental and Natural Resources Engineering, Luleå University of Technology, Luleå, Sweden
e-mail: sadia.ilyas@ltu.se

Wallenberg Initiative Materials Science for Sustainability (WISE), Luleå University of Technology, Luleå, Sweden

N. Ilyas (✉)
Institute of Microbiology, University of Agriculture Faisalabad (UAF), Faisalabad, Pakistan
e-mail: n7.ilyas@gmail.com

P. Pathak et al. (eds.), *Plastic Footprint*, https://doi.org/10.1007/978-981-96-6362-0_3

1 Introduction

Plastic has become an essential part of daily life due to its affordability, strength, lightweight properties, low electrical and thermal conductivity, and versatility in molding into various shapes (Almeshal et al., 2020). Since its inception in 1950, plastic production has skyrocketed, with the United Nations Environment Programme (UNEP) reporting that global plastic production reached approximately 400 million metric tons (Mt) annually by 2021. Cumulatively, the production of virgin plastics has totaled 8,300 Mt. However, by 2015, of the 6,300 Mt of plastic waste generated, only 9% was recycled, 12% was incinerated, and a staggering 79% ended up in landfills or was released into the environment. If current trends in plastic production and waste management continue, it is projected that by 2050, around 12,000 Mt of plastic waste will accumulate in landfills or pollute the environment (Tota-Maharaj et al., 2022). Figure 1 presents the global generation, disposal, incineration, and recycling of plastic waste in millions of metric tons from 1950 to 2050, with data from 2015 to 2020 based on projections reflecting current consumption patterns (Geyer et al., 2017). Despite the numerous advantages of plastic, improper disposal poses significant environmental risks, including the release of hazardous chemicals and greenhouse gases (GHGs), which contribute to air pollution and accelerate climate change (Sharma & Bansal, 2016). Figure 2 illustrates the environmental impacts associated with plastic waste (Hamada et al., 2024).

Alternatively, the development and use of biodegradable plastics has been proposed as a potential solution to the growing issue of plastic waste (Khardenavis et al., 2007; Nonato et al., 2001). While this approach may help in addressing long-term challenges of plastic waste, but do not provide a comprehensive solution for the recycling of oceanic plastic debris. Ocean plastic waste is particularly difficult to collect and manage as compared to land-based waste, which is easier to sort and

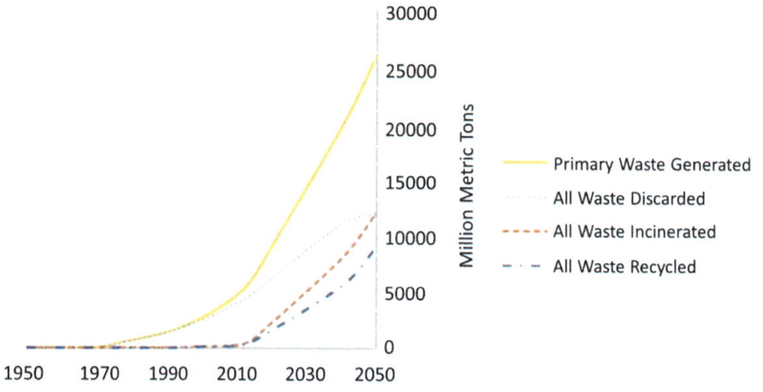

Fig. 1 Global cumulative plastic waste generation and disposal (*Source data* taken from Geyer et al., 2017; Tota-Maharaj et al., 2022)

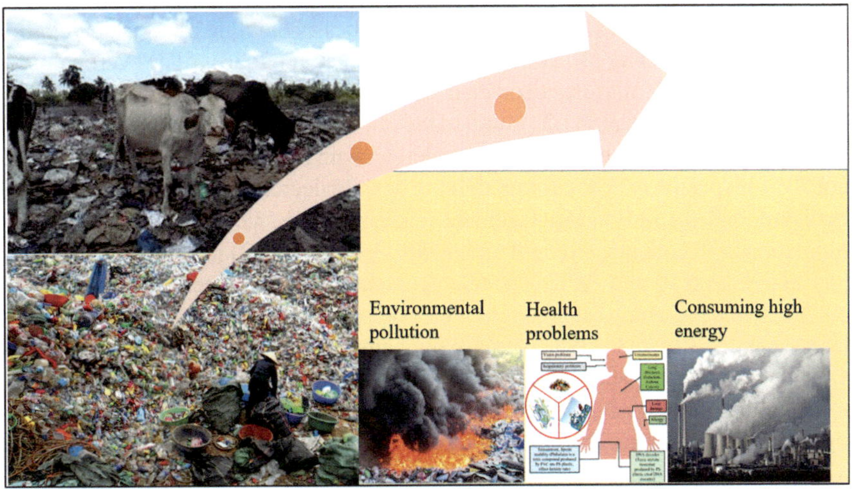

Fig. 2 Illustration of environmental issues associated with plastic waste disposal

process. Common plastics found in waste streams include low-density polyethylene (LDPE), high-density polyethylene (HDPE), linear low-density polyethylene (LLDPE), polypropylene (PP), nylon 6, polystyrene (PS), polyvinyl chloride (PVC), and polyethylene terephthalate (PET) (Hopewell et al., 2009).

Various methods are employed to manage plastic waste, each with its own benefits and drawbacks. Conventional approaches, such as landfilling, are cost-effective but pose long-term environmental risks, including leachate contamination and slow decomposition (Lin et al., 2023). Incineration, while reducing waste volume and generating energy, also releases pollutants and produces ash (Verma et al., 2016). Pyrolysis transforms plastic into fuel but requires significant energy input (Qureshi et al., 2020), while chemical recycling produces high-quality raw materials, albeit it is expensive and energy intensive (Huang et al., 2022). Recycling plastic waste into valuable products is a more efficient approach, addressing disposal issues and increasing the economic value of the waste (Khalid et al., 2022; Shamsuyeva & Endres, 2021). Further, reusing plastic waste (PW), such as incorporating it into concrete or utilizing additional recycling methods to create useful products, has a lower environmental impacts compared to incineration or burning (Almeshal et al., 2020; Almohana et al., 2022). Concrete, the second-most consumed material after water, is used at a rate of approximately three tons per person annually and is utilized twice as frequently as all other building materials combined in construction. Thus, concrete presents a highly effective means of reducing plastic pollution in ecosystems, infrastructure, and across water and land resources (Tota-Maharaj et al., 2022).

Plastic waste has been utilized in various studies as building materials, including aggregates (Al-Tayeb et al., 2022; Belmokaddem et al., 2020), filler (Veropalumbo et al., 2021), asphalt mixtures (You et al., 2022), insulation (Zulkernain et al., 2021), and more. Kangavar et al. (2022) studied the effects of using polyethylene terephthalate (PET) waste as a fine aggregate in concrete, with replacement levels of 0, 10, 30, and 50%, respectively. They found that the density decreased as the amount of PET increased, and the optimal replacement level was 10%, which resulted in the best improvement in concrete strength and ductility. Islam (2022) investigated the use of polypropylene (PP) and PET as coarse aggregates at replacement levels of 10, 20, and 30%. It was found that concrete with polypropylene (PP) aggregates had a 9% decrease in density and a 39% increase in compressive strength. However, concrete with PET aggregates showed a 53% reduction in compressive strength compared to the control sample. Bachtiar et al. (2022) examined the use of PET as a partial replacement for coarse aggregates and found that adding PET significantly improved both the workability and strength of the concrete.

Therefore, this chapter provides a comprehensive review on recycling PW materials and their integration into concrete production. It explores various methods for recycling plastic waste and examines their incorporation into cement-based materials. The study focuses on the impact of recycled plastics based on the physical and mechanical properties of these materials, such as strength, durability, workability, and density. By summarizing key findings from existing literature, this work aims to contribute to a deeper understanding of how recycled plastic can be effectively used in cement-based materials to achieve both environmental and engineering applications.

2 Commonly Used Recycled Materials in Construction

PW is a significant contributor to the construction industry's consumption of various waste materials. Utilizing this waste in construction can improve sustainability and help to address the environmental issues caused by its accumulation in landfills and open spaces (Dhawan et al., 2019). Recycled materials are commonly used in construction as fibers and aggregates to enhance the tensile strength of cement-based products. Recycled PET is often used as coarse aggregate in concrete, fine aggregate in mortar, and recycled PS as lightweight aggregates in concrete. Additionally, HDPE, nylon 6, and recycled PET are used as reinforced material in cementitious composites, however, studies on this topic remain limited (Park & Kim, 2020).

2.1 Aggregates

Numerous studies have examined the physical properties of plastic waste to determine its potential use as fine or coarse aggregates (Belmokaddem et al., 2020; Olofinnade et al., 2021). Table 1 summarizes examples of recycled plastics and using them as aggregates. While various plastics such as HDPE (Almohana et al., 2022), PET (Kangavar et al., 2022), LLDPE (Lin et al., 2023), LDPE, and PP are utilized in concrete sector, the majority of plastics employed are recycled PET bottles (Park & Kim, 2020).

Figure 3 illustrates the appearance of recycled plastic. The primary method for converting plastic into aggregate is shredding, which involves pulverizing the plastic. Some studies have also used pellets of specific sizes made by melting the plastic (Ferreira et al., 2012). Additionally, crushing and grinding techniques have been explored, primarily to regulate the recycled material size, and these methods are largely similar to grinding (Park & Kim, 2020).

2.2 Recycled Plastics as Reinforcing Fibers in Cement-Based Materials

Table 2 highlights studies using PET, nylon 6, PP, HDPE, and GFRP as reinforcing fibers in cement-based materials (Park & Kim, 2020). Synthetic fibers, including PET, have been shown to improve concrete's toughness and crack resistance (Park, 2011). Foti (2011) found that PET bottle fibers offer tensile strength comparable to carbon or steel fibers. PP/PE-blended fiber-reinforced composites (HyFRCs) at 2.9% fiber volume improved compressive and flexural strengths by 38 $\pm$ 2% and 40 $\pm$ 1%, respectively (Silva et al., 2013). Morphological studies revealed strong mechanical interactions between the fibers and the cement matrix, similar to those in polyacrylonitrile-reinforced composites (PANFRCs) (Khalid et al., 2018).

Many studies have found that fiber pullout strength increases with the fiber's embedded length in the concrete matrix (Beglarigale & Yazici, 2015). Richardson and Landless (2009) reported a 39.3–48.1% increase in fiber strength when the embedded length ranged from 45 to 55 mm. This increase is due to the surface area of the fiber interacting with the concrete, which influences friction and bond strength at the interface. The fiber-concrete contact area influences both pullout resistant and interfacial bond energy (Alberti et al., 2014; Babafemi & Boshoff, 2015; Yoo et al., 2013). Vairagade et al. (2012) used PP fibers of lengths 15, 20, and 24 mm with 0.4% fiber content, finding strength increases of 1.2 to 4.5%. Longer fibers have a larger surface area, leading to stronger interfacial bond strength and higher friction energy under load. Thus, fiber length significantly impacts strength. Table 3 summarizes the properties of recycled fibers used in existing studies (Khalid et al., 2018; Park & Kim, 2020).

Table 1 Type of plastic aggregate and target materials in the literature

Type of composites	Origin of plastic waste	Type of plastic	Recycling procedure	Target	References
Concrete	Mixed	HDPE	Shredding	Fine aggregate	Naik et al. (1996)
Lightweight concrete	–	EPS	Granular	Coarse aggregate	Sabaa and Ravindrarajah (1997)
Concrete	Bottle	PET	Grinding	Fine aggregate	Choi et al. (2005)
Non-load-bearing lightweight concrete	Form	PS	Crumbling	Fine aggregate	Laukaitis et al. (2005)
Concrete	Mixed	Plastic	Grinding	Fine aggregate	Batayneh et al. (2007)
Mortar	Bottle	PET	Shredding	Fine aggregate	Marzouk et al. (2007)
Non-load-bearing lightweight concrete	Melamine	Melamine	Grinding	Fine aggregate	Panyakapo and Panyakapo (2008)
Concrete	–	Mix plastic	Crushing	Fine aggregate	Ismail and Al-Hashmi (2008)
Lightweight concrete	EPS form	Modified EPS	Thermal treatment	Coarse aggregate	Kan and Demirboğa (2009)
Concrete	–	GFRP	Grinding	Fine aggregate	Asokan et al. (2009)
Mortar and concrete	Bottle	PET	Mixed with river	Fine aggregate	Choi et al. (2009)
Concrete	Bottle	PET	Grinding	Fine aggregate	Albano et al. (2009)
Lightweight concrete	Pipe	PVC	Crushing	Fine aggregate	Kou et al. (2009)
Mortar	Bottle	PET	Shredding	Fine aggregate	Remadnia et al. (2009)
Lightweight concrete	Bottle	PET	Shredding	Fine aggregate	Akçaözoğlu et al. (2010)
Concrete	–	GFRP	Grinding	Fine aggregate	Asokan et al. (2010)
Lightweight concrete	Sheet	EVA	Crushing	Coarse aggregate	Lima et al. (2010)
Concrete	Bottle	PET	-	Fine aggregate	Frigione (2010)

(continued)

Table 1 (continued)

Type of composites	Origin of plastic waste	Type of plastic	Recycling procedure	Target	References
Lightweight concrete	Form	PUR	Crushing	Coarse aggregate	Fraj et al. (2010)
Concrete	Bag, bottle	PET and LDPE	Crushing and agglutinating	Fine aggregate	Galvão et al. (2011)

(a) (b) (c) (d)

Fig. 3 Typical representation of recycled plastic aggregates like **a** Shredded PET, **b** PET coarse aggregate, **c** PP aggregates, and **d** LDPE pellets

2.3 *Waste Fishing Nets for Sustainable Construction Applications*

Fishing nets are made from durable, non-biodegradable materials like nylon, HDPE, PET, and polypropylene, which are recycled into various products such as furniture, accessories, and recreational items. In civil engineering, waste fishing nets are used as fiber reinforcement in soil, mortar, and concrete to enhance strength and ductility. They also improve abrasion resistance in bituminous roads and provide resistance to tensile and impact stresses when added to cementitious composites, offering a sustainable alternative to synthetic fibers (Gopinath et al., 2024) (Table 4).

3 Properties of Construction Materials with Recycled Plastics Integration

To review previous research and assess the impact of plastic recycling on cement-based materials, the relative characteristics were calculated as

$$\text{Relative property} = \frac{\textit{Test result with plastic}}{\textit{Test results without plastic}} \quad (1)$$

Table 2 Types of plastic fibers in the literature

Origin of plastic waste	Type of plastic	Recycling procedure	Type of composites	References
Carpet	PP and Nylon 6	Grinding	Concrete	Wang et al. (1994)
Bottle	PET	Extruding	Mortar	Silva et al. (2005)
Bottle	PET	Extruding	Concrete	Ochi et al. (2007)
Bottle	PET	Melting and slitting	Mortar	Kim et al. (2008)
Bag	PP	Cutting	Concrete	Meddah and Bencheikh (2009)
Bottle	PET	Melting and slitting	Concrete	Kim et al. (2010)
Bottle	PET	Cutting	Mortar	Oliveira and Castro-Gomes (2011)
Bottle	PET	Extruding	Concrete	Fraternali et al. (2011)
Bottle	PET	Extruding	Concrete	Pelisser et al. (2012)
Bottle	PET	Cutting	Concrete	Nibudey et al. (2013)
Pot	Plastic	Cutting	Concrete	Prahallada and Parkash (2013)
Bottle	PET	Extruding	Concrete	Fraternali et al. (2014)
–	GFRP	Milling	Concrete	García et al. (2014)
Waste fishing net	Nylon 6	Cutting	Mortar	Spadea et al. (2015)
Waste fishing net	Nylon 6	Cutting	Mortar	Orasutthikul et al. (2017)
Waste fishing net	HDPE	Cutting	Mortar	Park et al. (2021)

The study assessed mechanical and physical properties, including workability, density, compressive strength, flexural strength, and split tensile strength. Recycled plastic fibers were added as a volume content in the test specimens, while recycled plastic aggregates replaced the original aggregates by volume or weight.

Table 3 Properties of recycled plastic fiber in the literature

Geometry of fiber	Diameter (mm)	Length (mm)	Aspect ratio	Tensile strength (MPa)	References
–	–	3–25	–	–	Wang et al. (1994)
Straight	0.026	20	769	323.5	Silva et al. (2005)
Straight	0.7	30	43	450	Ochi et al. (2007)
Straight Crimped Embossed	0.5 × 1.0 0.3 × 1.2 0.2 × 1.3	50	87	–	Kim et al. (2008)
Flat	–	60	–	–	Meddah and Bencheikh (2009)
Embossed	0.2 × 1.3	50	87	−420.7	Kim et al. (2010)
Flat	0.5 × 0.2	35	31	–	Oliveira and Castro-Gomes (2011)
Straight Straight Crimped	1.1 0.7 0.7	40 52 52	36 74 74	550 263.72 274.29	Fraternali et al. (2011)
Multifilaments	0.025	20	800	–	Pelisser et al. (2012)
Flat	0.71	25	35	–	Nibudey et al. (2013)
–	–	–	30–110	–	Prahallada and Parkash (2013)
Straight Crimped	1.1 0.7	40 52	36 74	550 274	Fraternali et al. (2014)
Fluff	–	–	–	–	García et al. (2014)
–	–	–	15–60	–	Sharma and Bansal (2016)
Straight Straight Straight	0.33 0.33 0.33	12.7 25.4 38.1	38 77 115	319 342 389	Spadea et al. (2015)
Straight Straight Knotted	0.35 0.35 0.35	20 40 40	57 86 114	440 440 440	Orasutthikul et al. (2017)

(continued)

Table 3 (continued)

Geometry of fiber	Diameter (mm)	Length (mm)	Aspect ratio	Tensile strength (MPa)	References
Twisted Multifilaments Twisted Multifilaments	1.2 1.2	20 40	17 33	260 260	Ottosen et al. (2019)
Twisted Multifilaments Twisted Multifilaments	1 1.5	40 40	40 27	188 173	Park et al. (2021)

Table 4 Summary of the waste fishing net in construction projects

Origin of plastic waste	Plastic type	Composite type	Recycling process	References
Waste fishing net	HDPE	Light weight soil	Reinforcing textile	Kim et al. (2008)
Waste fishing net	HDPE	Mortar	Reinforcing textile	Mousa (2017)
Waste fishing net	–	Concrete	Reinforcing bar	Bertelsen and Ottosen (2016)
Waste fishing net	–	Concrete	Reinforcing bar	Sigvardsen (2016)

3.1 Physical Properties

3.1.1 Effect of Recycled Plastic Aggregate

The use of plastic waste as fine and coarse aggregates in concrete mixtures has been extensively researched, demonstrating a significant impact on concrete properties (Al-Tayeb et al., 2022; Liu et al., 2022; Lu et al., 2022). Various types of plastic waste, including PP, PVC, PET, LDPE, HDPE, and EPS are used to reduce concrete density due to their lightweight and low-density characteristics compared to natural aggregates (Hamada et al., 2024). The workability of cement-based materials is illustrated in Fig. 4, which also demonstrates the relationship between slump and slump flow for lightweight concrete, mortar, and concrete with different percentages of plastic aggregate replacement. Most studies, except those by Choi et al. (2005) and Choi et al. (2009), show that workability decreases as plastic aggregate content increases. Albano et al. (2009) explain that plastic aggregates absorb less water (0.1%) than cement aggregates (1.83%), which leads to increased porosity and reduced workability. Saikia and De Brito (2014) found that pellet-shaped aggregates with smooth surfaces improved workability, while irregularly shaped aggregates (crushed or shredded) reduced it. Additionally, Choi et al. (2009) found that mixing plastic with river sand to form smooth, spherical aggregates reduced friction and

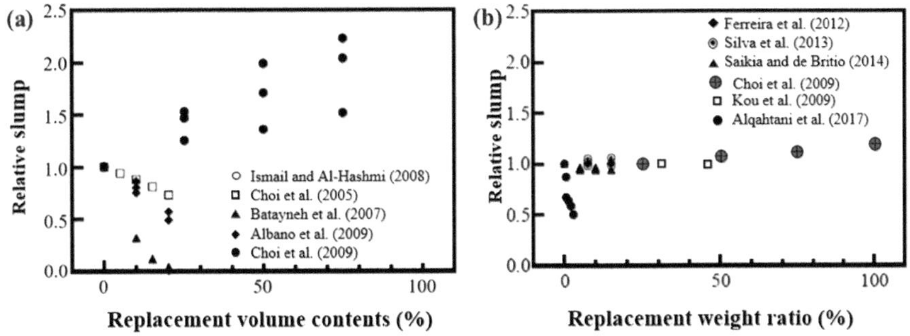

Fig. 4 Relative workability with plastic aggregates: **a** replaced volume in concrete; **b** replaced weight in concrete; **c** replaced volume in mortar; **d** replaced volume in lightweight concrete [*Figure Source* Park & Kim, 2020 published under open access license to use, wherein the figures were redrawn from the Choi et al. (2005), Choi et al. (2009), Ismail and Al-Hashmi (2008), Batayneh et al. (2007), Kou et al. (2009), Albano et al. (2009), Silva et al. (2005), Ferreira et al. (2012), Silva et al. (2013), Saikia and De Brito (2014), Alqahtani et al. (2017)]

improved workability. In order to increase the workability of cement-based products that are substituted with plastic aggregates, the shape of the aggregate is crucial (Park & Kim, 2020).

Figure 5 shows how the initial density of concrete is influenced by the replacement of aggregates with plastic. As the plastic aggregate content increased, the initial density decreased, with the extent of reduction varying depending on the replacement method. Ismail and Al-Hashmi (2008) found that replacing fine aggregates with PE aggregates (density 386.7 kg/m^3) reduced the initial density by about 93%. The reduction was less when the fine aggregates were replaced by weight rather than by volume. Badache et al. (2018) observed a smaller decrease in initial density when HDPE (density 922 kg/m^3) was used as fine aggregate. Figure 6d indicates that replacing coarse aggregates in lightweight concrete resulted in a smaller reduction in initial density compared to replacing fine aggregates. Alqahtani et al. (2017) reported a slight increase in initial density when substituting a lightweight aggregate with a density of 1410 kg/m^3 with a plastic aggregate having a density of 1540 kg/m^3 (Park & Kim, 2020).

3.1.2 Effect of Recycled Plastic Fibers

Using recycled plastic as fibers reduces workability and density similar to its use as aggregates. However, fiber geometry improves shrinkage resistance. Concrete with recycled plastic aggregates generally exhibits higher shrinkage due to the low stiffness of the plastic, but shrinkage decreases when fibers are used (Akçaözoğlu et al., 2010; Asokan et al., 2010). Fibers resist tensile forces caused by shrinkage through interfacial shear resistance with the matrix. Increased fiber volume reduces crack width and total crack area from plastic shrinkage (Auchey, 1998; Bertelsen

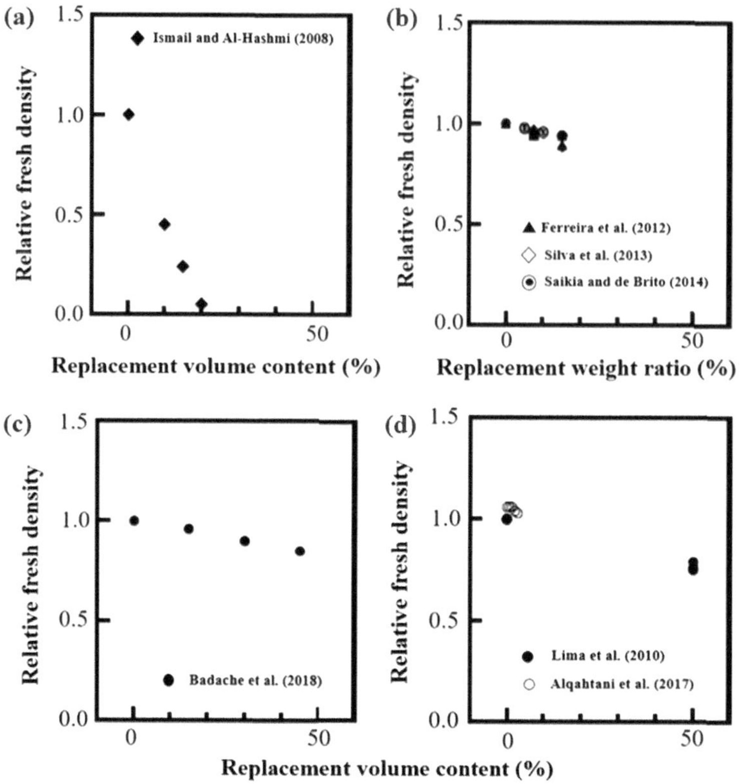

Fig. 5 Relative fresh density with plastic aggregates: **a** replaced volume in concrete; **b** replaced weight in concrete; **c** replaced volume in mortar; **d** replaced volume in lightweight concrete [*Figure Source* Park & Kim, 2020 published under open access license to use, wherein the figures were redrawn from Ismail and Al-Hashmi (2008); Silva et al. (2005); Ferreira et al. (2012); Saikia and De Brito (2014); Alqahtani et al. (2017); Badache et al. (2018); Lima et al. (2010)]

et al., 2019; Kim et al., 2008; Kim et al., 2008). Kim et al. (2008) noted that fiber geometry affects shrinkage crack control, but this impact diminishes when the fiber volume ratio exceeds 0.5%.

Based on the volume composition of recycled plastic fibers, Fig. 6 shows the link between slump flow and toughened density. Figure 6a illustrates how slump flow diminishes as the amount of plastic fiber increases. According to Nibudey et al. (2013), this is because fibers prevent aggregate movement, which impacts the workability of fiber-reinforced concrete (FRC). According to Orasutthikul et al. (2017), slump flow is impacted by fiber form because of fiber balling, which results in fiber-mortar separation. As a result, adding recycled plastic reduces workability. When fiber shape tends to cause agglomeration, slump flow is higher for the same fiber content compared to well-dispersed fibers, due to fiber-mortar separation. As shown in Fig. 6b, the hardened density of FRC decreases as the volume content of plastic

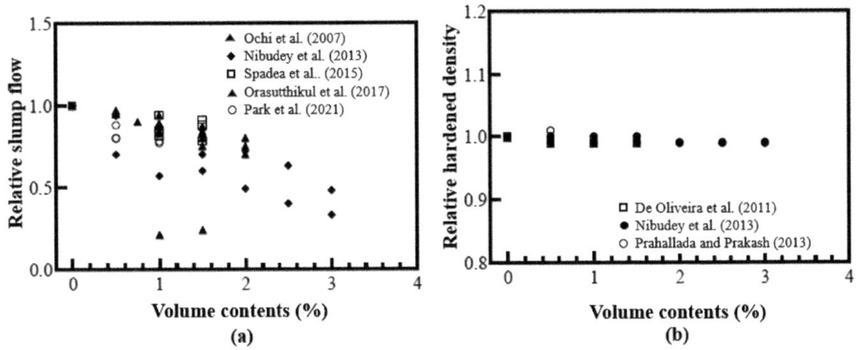

Fig. 6 Relative physical properties of fiber-reinforced concrete (FRC) with recycled plastic fiber: **a** slump flow; **b** hardened density [*Figure Source* Park & Kim, 2020 published under open access license to use, wherein the figures were redrawn from Ochi et al. (2007); Nibudey et al. (2013); Oliveira and Castro-Gomes (2011); Spadea et al. (2015); Prahallada and Parkash (2013); Orasutthikul et al. (2017); Park et al. (2021)]

fibers increases. However, the reduction in hardened density is minimal compared to conventional concrete (without plastic fibers), as the fiber content is relatively low (0.5–3.0%) (Park & Kim, 2020).

3.2 Mechanical Properties

The effect of PW aggregates on the mechanical properties of concrete has been the subject of numerous investigations. According to these investigations, concrete and mortar's compressive strength decreases with increased plastic waste levels (Hamada et al., 2024). Interestingly, concrete with less than 5% plastic aggregate by weight sometimes exhibits higher compressive strength than conventional concrete, though the reasons remain unclear. Some researchers suggest an optimal replacement level of around 2% by weight (Asokan et al., 2010; Fraj et al., 2010). Marzouk et al. (2007) analyzed mortar microstructure using SEM and found that PET aggregate replacement up to 50% maintained high compressibility. However, they also noted a significant increase in porosity when the replacement exceeded 50%.

Tensile strength is crucial for assessing the ductility of concrete materials. Studies show that using PW as aggregate typically reduces the splitting tensile strength of concrete (Hossain et al., 2016). This reduction is primarily attributed to weaker adhesion between the plastic aggregates and the cement matrix, which becomes more pronounced as the plastic aggregate replacement ratio increases (Albano et al., 2009; Ferreira et al., 2012; Kou et al., 2009; Saikia & Brito, 2013). Saikia and De Brito (2014) further explained that the smooth surface of plastic aggregates and the presence of free water on their surface weaken the bond with the cement paste. Replacing fine aggregates with plastic at the same water–cement ratio increases porosity and reduces tensile strength due to decreased water absorption. Flexural strength, or modulus of rupture, measures concrete's ability to resist bending under high loads. The inclusion of PW as aggregate affects the flexural strength of concrete and cement mortars. Similar to compressive and splitting tensile strength, flexural strength tends to decrease as the content of plastic aggregate increases (Park & Kim, 2020).

3.2.1 Effect of Recycled Plastic Fibers

The recycled plastic fibers significantly influence the mechanical properties of fiber-reinforced concrete (FRC). When the volume ratio exceeded 2%, compressive strength decreased compared to fiber-free cement-based materials. However, up to 2%, the mechanical properties varied based on fiber shape. Fraternali et al. (2011) found that 1% recycled PET fibers increased compressive strength when the fiber density was higher than that of conventional concrete. Spadea et al. (2015) noted a decrease in strength due to the deformability of plastic fibers, which create voids under compressive force. Prahallada and Parkash (2013) reported that increasing the fiber aspect ratio above 50 reduced compressive strength at the same volume. Therefore, fiber density and aspect ratio influence the compressive strength of recycled plastic FRC.

The split tensile strength and flexural strength increased with fiber volume content up to 1% and 1.5%, respectively. However, increasing the fiber volume from 1.5% to 3.0% led to a decrease in flexural strength (Nibudey et al., 2013; Silva et al., 2005). Orasutthikul et al. (2017) identified a balling issue when higher fiber volumes were used or when fibers were poorly distributed, which negatively impacted flexural strength. To determine the optimal fiber aspect ratio and volume, the reinforced index was calculated as

$$\text{Reinforced index} = \frac{Vf.Lf}{Df} = Vf \tag{2}$$

where *Vf*, *Lf*, *Df*, and *Lf/Df* are the fiber volume content, fiber length, fiber diameter, and aspect ratio. The relationship between the split tensile strength and flexural strength is shown in Fig. 7.

Yu et al. (2018) discovered that while the fiber-reinforced index increased the tensile strength of FRC, it had no discernible effect on the split tensile and flexural strengths of FRC made from recycled plastic. To find out why reinforced plastic fiber reinforcement has no effect on the reinforced index and how to best fabricate recycled plastic fibers to increase tensile strength, more research is required. The results of a direct tensile test employing recycled fishing nets by Park et al. (2021) were found to be insufficient, indicating the necessity of more direct tensile testing on recycled plastic fibers in subsequent research.

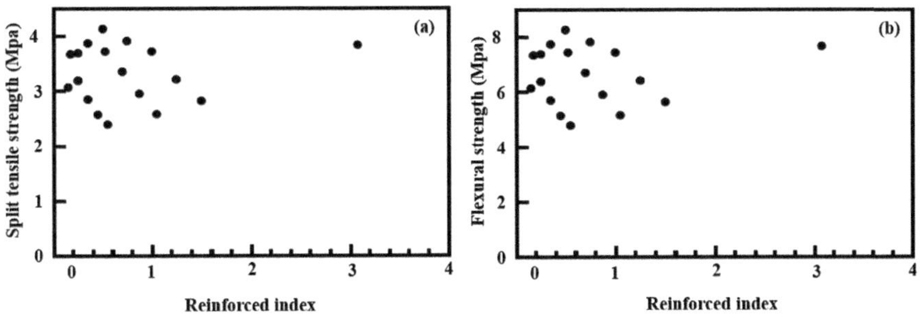

Fig. 7 Effect of fiber-reinforced index: **a** on split tensile strength; **b** on flexural strength [*Figure Source* Park & Kim, 2020 published under open access license to use]

3.3 Comparison of Properties Between Virgin and Recycled Plastics for Construction Materials

In these studies where virgin plastic was used as an aggregate, Expanded Polystyrene (EPS) was commonly employed to create lightweight concrete. However, the use of virgin plastic for aggregates has decreased in recent years due to the challenges it shares with recycled plastic aggregates, particularly the reduction in strength caused by a decrease in density. A study by Alqahtani et al. (2017) showed that concrete made with recycled Low-Density Polyethylene (LLDPE) aggregates, replacing volcanic lightweight coarse aggregates, had increased density but experienced significant reductions in compressive strength (15–62%), flexural strength (27–44%), and splitting tensile strength (12–31%). As a result, recycled plastic aggregates for lightweight concrete are still not widely used in construction.

In contrast, virgin plastic fibers have been more widely applied in specialized structures such as high-rise buildings, protection facilities, tunnel linings, foundations, and concrete pavements. PP is a common virgin plastic material used to enhance the tensile strength of cement-based materials, while PET and PE are more frequently recycled for construction applications. Although both virgin and recycled plastic fibers are made from the same base materials, recycled fibers typically show less reduction in tensile strength compared to their virgin counterparts after the recycling process. Several studies have explored the mechanical behavior of both virgin and recycled plastic fibers in cement-based materials (Park & Kim, 2020). For example, Kim et al. (2010) compared the performance of recycled PET fibers with commercial PP fibers in concrete. Their findings showed similar results at the same volume ratio, suggesting that recycled fibers could effectively replace virgin fibers in cement-based materials. Similarly, Fraternali et al. (2011) found that high-strength PET fibers significantly improved compressive and flexural strength compared to both unreinforced and PP fiber-reinforced concrete. They also noted that corrugated recycled PET fibers enhanced ductility, which is important for improving concrete's flexibility and crack resistance. Orasutthikul et al. (2017) conducted additional research that compared the reinforcing effects of virgin PVC fibers, recycled nylon, and PET.

They found that compared to virgin PVC fibers, embossed recycled PET fibers offered greater ductility and compressive strength.

These findings suggest that recycled plastic fibers, including those sourced from ocean plastics, can be effectively used to reinforce cement-based materials, provided they undergo appropriate pretreatment. This opens the possibility of using ocean plastic waste in construction, reducing the need for virgin plastics and promoting environmental sustainability. In conclusion, while recycled plastic aggregates still face challenges in terms of strength and density, recycled plastic fibers, particularly those from ocean plastics, offer significant potential. With further research into fiber treatment and optimization, recycled plastic fibers could become a valuable resource for enhancing the mechanical properties of cement-based materials, contributing to both sustainability and plastic waste reduction.

3.4 Recent Advancements in Terms of Machine Learning for Predicting the Mechanical Properties

In summary, the plastic waste can be effectively used as a suitable alternative to concrete materials; however, the studies conducted via experimental approach are tedious time-consuming work. Moreover, the concrete properties are greatly affected by the factors like additives proportion, etc. Hence, to establish a strong correlation between the mechanical properties of plastic concrete and other ingredients mixed, and exhibited nonlinear behavior between them, recently, a few works are reported to apply the artificial intelligence (AI) in this field. Asif et al. (2024a) performed a comparative analysis of individual and ensemble learning algorithms (ELAs), predicting the compressive strength (CS) and flexural strength (FS) of plastic concrete (as the conceptual diagram depicted in Fig. 8).

They utilized a combination of multi-layer perceptron neural network (MLPNN), random forest (RF), support vector machine (SVM), and decision tree (DT) as base learners, and enhanced their predictive performance by integrating them with bagging and AdaBoost methods. To predict the CS and FS of plastic concrete, computational models were developed using gene expression programming (GEP). Based on statistical metrics and other validation criteria, GEP outperformed all individual models. However, for extreme learning algorithms (ELAs), the SVR ensemble combined with AdaBoost and the RF model modified with bagging achieved superior performance. Sensitivity analysis and SHapley Additive exPlanations (SHAP) identified that cement, plastic, water, and specimen age had the greatest role on predictions, while the impact of superplasticizer was found to be the least significant.

Furthermore, Asif et al. (2024b) have compared the multi-expression programming (MEP) with the multi-linear regression (MLR) model as shown in Fig. 9. A sensitivity analysis was performed to determine the contribution of each parameter to CS and tensile strength (TS) of plastic concrete. The statistical results showed that the MEP model demonstrated higher efficacy in prediction, with an R^2-value ≥ 0.87

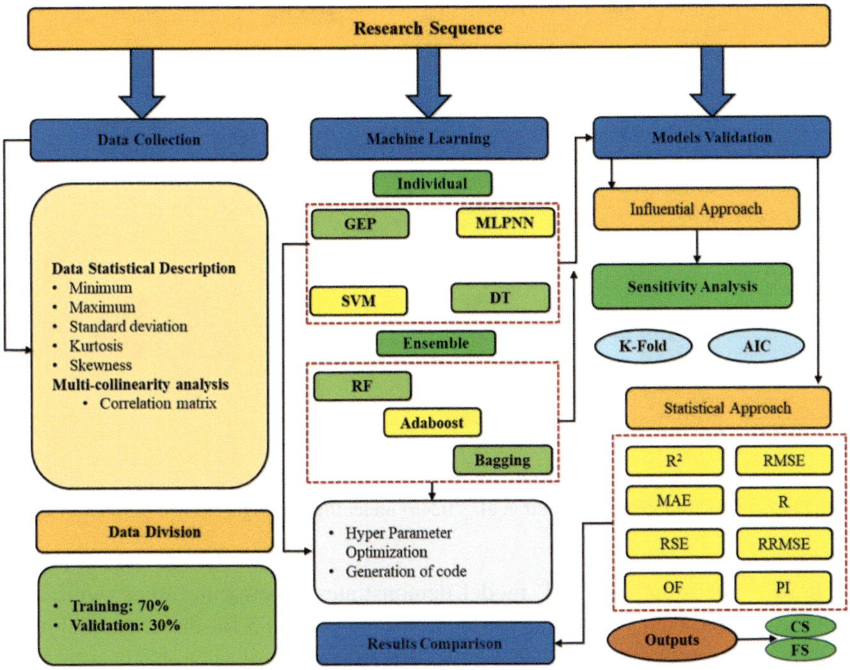

Fig. 8 Schematic flow diagram depicting the sequence of research using AI in predicting the mechanical properties of plastic concrete (adopted from Asif et al. (2024a) under the copyright license from Elsevier)

for both models of CS and TS, as follows:

$$\mathrm{CS(MPa)} = \left[\frac{-3\left(\mathrm{P} + 2\mathrm{C}\left(\frac{c^2}{w^3} + \frac{c}{w^2}\right) + \mathrm{A}\left(\mathrm{A} + \mathrm{C}\left(\frac{c^2}{w^3} + \frac{c}{w^2}\right)\right)^2\left(\frac{c^2}{w^3} + \frac{c}{w^2}\right)\right)^2}{C} - \frac{C^2\left(\mathrm{P} - \mathrm{A}\left(\mathrm{A} + \mathrm{C}\left(\frac{c^2}{w^3} + \frac{c}{w^2}\right)\right)^2\left(\frac{c^2}{w^3} + \frac{c}{w^2}\right)^2\right)}{W^3} + \frac{\left(\mathrm{A} + \mathrm{C}\left(\frac{c^2}{w^3} + \frac{c}{w^2}\right) + \frac{c}{w}\right)\mathrm{W}}{C} + \frac{\mathrm{S}\left(\mathrm{A} + \mathrm{C}\left(\frac{c^2}{w^3} + \frac{c}{w^2}\right) + \frac{c}{w}\right)\mathrm{W}}{\mathrm{AC}\left(\mathrm{A} = \mathrm{G}\left(\frac{c^2}{w^3} + \frac{c}{w^2}\right)\right)\left(G + 2A\left(\frac{c^2}{w^3} + \frac{c}{w^2}\right)\right)^2} \right] \tag{3}$$

$$Ts(\mathrm{MPa}) = \frac{-\frac{S}{W} + 3W + \frac{(W+P)^2}{3W}}{\frac{S}{G} + W + P} + \frac{A}{W} + \frac{\left(\mathrm{A} + P\left(\frac{c}{w^3} + \frac{c}{w^2}\right) + \frac{c}{w}\right)\mathrm{W}}{C} \tag{4}$$

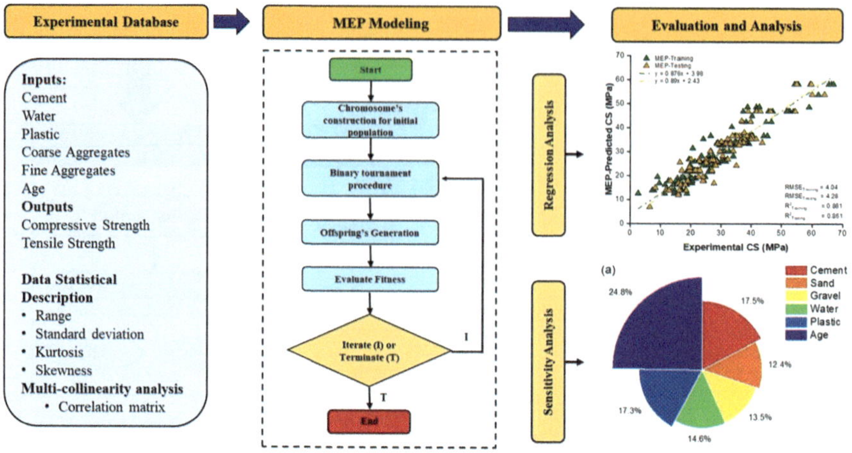

Fig. 9 Schematic diagram depicting the MEP versus MLR analysis, evaluating the concrete made with plastic waste (adopted from Asif et al., 2024b) under the copyright license from Elsevier)

It was noted that the MEP model demonstrates superior accuracy, with slope values 0.87 and 0.89 for the CS model and 0.88 and 0.885 for the TS model in the training and testing datasets, respectively. These results indicate a strong correlation between the predicted and observed values. Furthermore, the R^2 values for both models were 0.88 and 0.86 during training and 0.89 and 0.88 during testing, with the results being highly consistent. This reflects an excellent fit for the datasets and a well-trained model. In comparison to simpler models like basic linear regression, the MLR model used for estimating the CS and TS of plastic concrete incorporates multiple explanatory factors. The corresponding mathematical expressions for the MLR model are as follows:

$$\text{CS(MPa)} = [15.39 + 0.079\text{C} + 0.009\text{S}\ 0.0034\text{G}\ 0.137\text{W}\ + 0.282\text{P}\ + 0.461\text{A}] \tag{5}$$

$$\text{TS(MPa)} = [0.48 + 0.0043\text{C} + 0.001\text{S}\ 0.0002\text{G}\ 0.00341\text{W} + 0.0021\text{P} + 0.0391\text{A}] \tag{6}$$

It was observed that the MEP model resulted in a 43% and 49% reduction in MAE and RMSE values for the CS models, respectively. However, for the TS model, the MEP approach demonstrated reductions of 38.4% and 28.6% in MAE and RMSE compared to the MLR models. Additionally, the maximum errors in the MEP models were lower than those in the MLR models, primarily due to the highly nonlinear nature of concrete properties, which limits the practical application of traditional regression methods for predictive modeling tasks. Therefore, the proposed model application is possible, as it effectively captures the nonlinear behavior of plastic concrete and enhances performance prediction. These findings suggest that the approach is highly relevant for sustainable construction practices.

4 Conclusion

This chapter focuses on the use of cement-based materials incorporating recycled plastic waste in concrete. It emphasizes that plastic waste (PW) can enhance workability; reduce density; and lower compressive, tensile, and flexural strengths to acceptable levels. Recycled plastics are employed both as aggregates to reduce material weight and as fibers to improve tensile strength. However, these materials can often reduce workability and increase density. To mitigate these challenges, effective surface treatments and optimal fiber volume ratios are crucial. The inclusion of recycled plastic aggregates tends to decrease compressive, tensile, and flexural strengths due to increased porosity and reduced density. High-density plastics are recommended for aggregates, although their use is primarily beneficial for lightweight concrete. Recycled plastic fibers can lower compressive strength but improve shrinkage resistance and tensile strength, with an optimal fiber volume of less than 1.5% to avoid issues such as fiber balling. Interestingly, recycled fibers often perform better than virgin fibers in concrete applications. Additionally, the discussion on machine learning (ML) and artificial intelligence (AI) for predicting the mechanical properties of plastic-infused concrete, along with comparisons of multilayer perceptron (MLP) and multiple linear regression (MLR) analyses, suggests that these approaches can be applied to sustainable construction practices, contributing to environmental pollution reduction.

Acknowledgements The authors, M. Farhan and S. Ilyas, greatly acknowledge the Wallenberg Initiative Materials Science for Sustainability (WISE) funded by the Knut and Alice Wallenberg Foundation, for supporting this work.

References

Akçaözoğlu, S., Atiş, C. D., & Akçaözoğlu, K. (2010). An investigation on the use of shredded waste PET bottles as aggregate in lightweight concrete. *Waste Management (Oxford), 30*(2), 285–290.

Albano, C., Camacho, N., Hernández, M., Matheus, A., & Gutierrez, A. (2009). Influence of content and particle size of waste pet bottles on concrete behavior at different w/c ratios. *Waste Management (Oxford), 29*(10), 2707–2716.

Alberti, M., Enfedaque, A., & Gálvez, J. (2014). On the mechanical properties and fracture behavior of polyolefin fiber-reinforced self-compacting concrete. *Construction and Building Materials, 55*, 274–288.

Almeshal, I., Tayeh, B. A., Alyousef, R., Alabduljabbar, H., & Mohamed, A. M. (2020). Eco-friendly concrete containing recycled plastic as partial replacement for sand. *Journal of Materials Research and Technology, 9*(3), 4631–4643.

Almohana, A. I., Abdulwahid, M. Y., Galobardes, I., Mushtaq, J., & Almojil, S. F. (2022). Producing sustainable concrete with plastic waste: A review. *Environmental Challenges, 9*, Article 100626.

Alqahtani, F. K., Ghataora, G., Khan, M. I., & Dirar, S. (2017). Novel lightweight concrete containing manufactured plastic aggregate. *Construction and Building Materials, 148*, 386–397.

Al-Tayeb, M. M., Aisheh, Y. I. A., Qaidi, S. M., & Tayeh, B. A. (2022). Experimental and simulation study on the impact resistance of concrete to replace high amounts of fine aggregate with plastic waste. *Case Studies in Construction Materials, 17*, Article e01324.

Asif, U., Javed, M. F., Abuhussain, M., Ali, M., Khan, W. A., & Mohamed, A. (2024a). Predicting the mechanical properties of plastic concrete: An optimization method by using genetic programming and ensemble learners. *Case Studies in Construction Materials, 20*, Article e03135.

Asif, U., Javed, M. F., Alyami, M., & Hammad, A. W. (2024b). Performance evaluation of concrete made with plastic waste using multi-expression programming. *Materials Today Communications, 39*, Article 108789.

Asokan, P., Osmani, M., & Price, A. D. (2009). Assessing the recycling potential of glass fibre reinforced plastic waste in concrete and cement composites. *Journal of Cleaner Production, 17*(9), 821–829.

Asokan, P., Osmani, M., & Price, A. D. (2010). Improvement of the mechanical properties of glass fibre reinforced plastic waste powder filled concrete. *Construction and Building Materials, 24*(4), 448–460.

Auchey, F. L. (1998). The use of recycled polymer fibers as secondary reinforcement in concrete structures. *Journal of Construction Education, 3*(2), 131–140.

Babafemi, A. J., & Boshoff, W. P. (2015). Tensile creep of macro-synthetic fibre reinforced concrete (MSFRC) under uni-axial tensile loading. *Cement and Concrete Composites, 55*, 62–69.

Bachtiar, E., Ihsan, M., Irfan, U. R., Setiawan, A., Makbul, R., & Setiawan, A. M., et al. (2022). Utilization of artificial coarse aggregate from polyethylene terephthalate plastic waste in concrete. *International Review of Civil Engineering (I. RE. CE), 13*(1), 19–30.

Badache, A., Benosman, A. S., Senhadji, Y., & Mouli, M. (2018). Thermo-physical and mechanical characteristics of sand-based lightweight composite mortars with recycled high-density polyethylene (HDPE). *Construction and Building Materials, 163*, 40–52.

Batayneh, M., Marie, I., & Asi, I. (2007). Use of selected waste materials in concrete mixes. *Waste Management (Oxford), 27*(12), 1870–1876.

Beglarigale, A., & Yazici, H. (2015). Pull-out behavior of steel fiber embedded in flowable RPC and ordinary mortar. *Construction and Building Materials, 75*, 255–265.

Belmokaddem, M., Mahi, A., Senhadji, Y., & Pekmezci, B. Y. (2020). Mechanical and physical properties and morphology of concrete containing plastic waste as aggregate. *Construction and Building Materials, 257*, Article 119559.

Bertelsen, I., Ottosen, L., & Fischer, G. (2019). Quantitative analysis of the influence of synthetic fibres on plastic shrinkage cracking using digital image correlation. *Construction and Building Materials, 199*, 124–137.

Bertelsen, I. M. G., & Ottosen, L. M. (2016). Engineering properties of fibres from waste fishing nets. *Paper presented at the International RILEM Conference on Materials, Systems and Structures in Civil Engineering.*

Choi W. Y., Moon, D. J., Kim, Y. J., & Lachemi, M. (2009). Characteristics of mortar and concrete containing fine aggregate manufactured from recycled waste polyethylene terephthalate bottles. *Construction and Building Materials, 23*(8), 2829–2835

Choi, Y.-W., Moon, D.-J., Chung, J.-S., & Cho, S.-K. (2005). Effects of waste PET bottles aggregate on the properties of concrete. *Cement and Concrete Research, 35*(4), 776–781.

Choudhary, R., Kumar, A., Murkute, K. (2018). Properties of waste polyethylene terephthalate (PET) modified asphalt mixes: dependence on PET size, PET content, and mixing process. *Periodica Polytechnica Civil Engineering, 62*(3), 685–693.

Dhawan, R., Bisht, B. M. S., Kumar, R., Kumari, S., & Dhawan, S. (2019). Recycling of plastic waste into tiles with reduced flammability and improved tensile strength. *Process Safety and Environment Protection, 124*, 299–307.

Ferreira, L., De Brito, J., & Saikia, N. (2012). Influence of curing conditions on the mechanical performance of concrete containing recycled plastic aggregate. *Construction and Building Materials, 36*, 196–204.

Foti, D. (2011). Preliminary analysis of concrete reinforced with waste bottles PET fibers. *Construction and Building Materials, 25*(4), 1906–1915.

Fraj, A. B., Kismi, M., & Mounanga, P. (2010). Valorization of coarse rigid polyurethane foam waste in lightweight aggregate concrete. *Construction and Building Materials, 24*(6), 1069–1077.

Fraternali, F., Ciancia, V., Chechile, R., Rizzano, G., Feo, L., & Incarnato, L. (2011). Experimental study of the thermo-mechanical properties of recycled PET fiber-reinforced concrete. *Composite Structures, 93*(9), 2368–2374.

Fraternali, F., Spadea, S., & Berardi, V. P. (2014). Effects of recycled PET fibres on the mechanical properties and seawater curing of Portland cement-based concretes. *Construction and Building Materials, 61*, 293–302.

Frigione, M. (2010). Recycling of PET bottles as fine aggregate in concrete. *Waste Management (Oxford), 30*(6), 1101–1106.

Galvão, J. C. A., Portella, K. F., Joukoski, A., Mendes, R., & Ferreira, E. S. (2011). Use of waste polymers in concrete for repair of dam hydraulic surfaces. *Construction and Building Materials, 25*(2), 1049–1055.

García, D., Vegas, I., & Cacho, I. (2014). Mechanical recycling of GFRP waste as short-fiber reinforcements in microconcrete. *Construction and Building Materials, 64*, 293–300.

Geyer, R., Jambeck, J. R., & Law, K. L. (2017). Production, use, and fate of all plastics ever made. *Science Advances, 3*(7), Article e1700782.

Gopinath, S., Kumar, N. M., Ganesan, I., Kumar, V. R. (2024). Use of waste fishing nets in cementitious applications.

Hamada, H. M., Al-Attar, A., Abed, F., Beddu, S., Humada, A. M., Majdi, A., et al. (2024). Enhancing sustainability in concrete construction: A comprehensive review of plastic waste as an aggregate material. *Sustainable Materials and Technologies, 40*, Article e00877.

Hopewell, J., Dvorak, R., & Kosior, E. (2009). Plastics recycling: Challenges and opportunities. *Philosophical Transactions of the Royal Society b: Biological Sciences, 364*(1526), 2115–2126.

Hossain, M., Bhowmik, P., & Shaad, K. (2016). Use of waste plastic aggregation in concrete as a constituent material. *Progressive Agriculture, 27*(3), 383–391.

Huang, J., Veksha, A., Chan, W. P., Giannis, A., & Lisak, G. (2022). Chemical recycling of plastic waste for sustainable material management: A prospective review on catalysts and processes. *Renewable and Sustainable Energy Reviews, 154*, Article 111866.

İpek, S., Diri, A., & Mermerdaş, K. (2021). Recycling the low-density polyethylene pellets in the pervious concrete production. *Journal of Material Cycles and Waste Management, 23*, 272–287.

Islam, M. J. (2022). Comparative study of concrete with polypropylene and polyethylene terephthalate waste plastic as partial replacement of coarse aggregate. *Advances in Civil Engineering, 2022*(1), 4928065.

Ismail, Z. Z., & Al-Hashmi, E. A. (2008). Use of waste plastic in concrete mixture as aggregate replacement. *Waste Management (Oxford), 28*(11), 2041–2047.

Kan, A., & Demirboğa, R. (2009). A novel material for lightweight concrete production. *Cement and Concrete Composites, 31*(7), 489–495.

Kangavar, M. E., Lokuge, W., Manalo, A., Karunasena, W., & Frigione, M. (2022). Investigation on the properties of concrete with recycled polyethylene terephthalate (PET) granules as fine aggregate replacement. *Case Studies in Construction Materials, 16*, Article e00934.

Khalid, F., Irwan, J., Ibrahim, M. W., Othman, N., & Shahidan, S. (2018). Performance of plastic wastes in fiber-reinforced concrete beams. *Construction and Building Materials, 183*, 451–464.

Khalid, M. Y., Arif, Z. U., Ahmed, W., & Arshad, H. (2022). Recent trends in recycling and reusing techniques of different plastic polymers and their composite materials. *Sustainable Materials and Technologies, 31*, Article e00382.

Khardenavis, A. A., Kumar, M. S., Mudliar, S. N., & Chakrabarti, T. (2007). Biotechnological conversion of agro-industrial wastewaters into biodegradable plastic, poly β-hydroxybutyrate. *Bioresource Technology, 98*(18), 3579–3584.

Kim, Y., Kim, H., & Lee, G. (2008). Mechanical behavior of lightweight soil reinforced with waste fishing net. *Geotextiles and Geomembranes, 26*(6), 512–518.

Kim, S. B., Yi, N. H., Kim, H. Y., Kim, J.-H.J., & Song, Y.-C. (2010). Material and structural performance evaluation of recycled PET fiber reinforced concrete. *Cement and Concrete Composites, 32*(3), 232–240.

Kim, J. J.-H., Park, C.-G., Lee, S.-W., Lee, S.-W., Won, J.-P. (2008). Effects of the geometry of recycled PET fiber reinforcement on shrinkage cracking of cement-based composites. *Composites Part B: Engineering, 39*(3), 442–450.

Kou, S., Lee, G., Poon, C. S., & Lai, W. L. (2009). Properties of lightweight aggregate concrete prepared with PVC granules derived from scraped PVC pipes. *Waste Management (Oxford), 29*(2), 621–628.

Laukaitis, A., Žurauskas, R., & Kerien, J. (2005). The effect of foam polystyrene granules on cement composite properties. *Cement and Concrete Composites, 27*(1), 41–47.

Lima, P. R. L., Leite, M. B., & Santiago, E. Q. R. (2010). Recycled lightweight concrete made from footwear industry waste and CDW. *Waste Management (Oxford), 30*(6), 1107–1113.

Lin, X., Zhang, S., Yang, S., Zhang, R., Shi, X., & Song, L. (2023). A landfill serves as a critical source of microplastic pollution and harbors diverse plastic biodegradation microbial species and enzymes: Study in large-scale landfills, China. *Journal of Hazardous Materials, 457*, Article 131676.

Liu, T., Nafees, A., Javed, M. F., Aslam, F., Alabduljabbar, H., Xiong, J.-J., et al. (2022). Comparative study of mechanical properties between irradiated and regular plastic waste as a replacement of cement and fine aggregate for manufacturing of green concrete. *Ain Shams Engineering Journal, 13*(2), Article 101563.

Lu, T., Li, J., Han, X., Liu, R., Cai, C., & Wang, H. (2022). Effect of blowing agent concentration of ultra-low density NBR/PVC composite foams on compressive behavior and energy-absorption characteristics. *Journal of Vinyl & Additive Technology, 28*(3), 640–648.

Marzouk, O. Y., Dheilly, R., & Queneudec, M. (2007). Valorization of post-consumer waste plastic in cementitious concrete composites. *Waste Management (Oxford), 27*(2), 310–318.

Meddah, M. S., & Bencheikh, M. (2009). Properties of concrete reinforced with different kinds of industrial waste fibre materials. *Construction and Building Materials, 23*(10), 3196–3205.

Mohammed, A. (2019). Mechanical strength of concrete with PVC aggregates. In *Use of recycled plastics in eco-efficient concrete* (pp. 115–135). Elsevier.

Mousa, M. A. (2017). Low-cost fishing net-reinforced cement matrix overlay for substandard concrete masonry in coastal areas.

Naik, T. R., Singh, S. S., Huber, C. O., & Brodersen, B. S. (1996). Use of post-consumer waste plastics in cement-based composites. *Cement and Concrete Research, 26*(10), 1489–1492.

Nibudey, R., Nagarnaik, P., Parbat, D., & Pande, A. (2013). Strength and fracture properties of post consumed waste plastic fiber reinforced concrete. *International Journal of Civil, Structural, Environmental and Infrastructure Engineering Research and Development, 3*(2), 9–16.

Nonato, R. v., Mantelatto, P., & Rossell, C. (2001). Integrated production of biodegradable plastic, sugar and ethanol. *Applied Microbiology and Biotechnology, 57*, 1–5.

Ochi, T., Okubo, S., & Fukui, K. (2007). Development of recycled PET fiber and its application as concrete-reinforcing fiber. *Cement and Concrete Composites, 29*(6), 448–455.

Oliveira, d. L. A. P., Castro-Gomes, J. P. (2011). Physical and mechanical behaviour of recycled PET fibre reinforced mortar. *Construction of Building Materials, 25*(4), 1712–1717.

Olofinnade, O., Chandra, S., & Chakraborty, P. (2021). Recycling of high impact polystyrene and low-density polyethylene plastic wastes in lightweight based concrete for sustainable construction. *Materials Today: Proceedings, 38*, 2151–2156.

Orasutthikul, S., Unno, D., & Yokota, H. (2017). Effectiveness of recycled nylon fiber from waste fishing net with respect to fiber reinforced mortar. *Construction and Building Materials, 146*, 594–602.

Ottosen, L. M., Svensson, S. J., & Bertelsen, I. M. (2019). Discarded nylon fishing nets as fibre reinforcement in cement mortar. *WIT Transactions on Ecology and the Environment, 231*, 245–256.

Panyakapo, P., & Panyakapo, M. (2008). Reuse of thermosetting plastic waste for lightweight concrete. *Waste Management (Oxford), 28*(9), 1581–1588.
Park, S.-S. (2011). Unconfined compressive strength and ductility of fiber-reinforced cemented sand. *Construction and Building Materials, 25*(2), 1134–1138.
Park, J. K., & Kim, M. O. (2020). Mechanical properties of cement-based materials with recycled plastic: A review. *Sustainability, 12*(21), 9060.
Park, J. K., Kim, D. J., & Kim, M. O. (2021). Mechanical behavior of waste fishing net fiber-reinforced cementitious composites subjected to direct tension. *Journal ofBuilding Engineering, 33*, Article 101622.
Pelisser, F., Montedo, O. R. K., Gleize, P. J. P., & Roman, H. R. (2012). Mechanical properties of recycled PET fibers in concrete. *Materials Research, 15*, 679–686.
Prahallada, M., & Parkash, K. (2013). Effect of different aspect ratio of waste plastic fibers on the properties of fiber reinforced concrete e an experimental investigation. *International Journal of Advanced Research in IT and Engineering, 2*, 1–13.
Qureshi, M. S., Oasmaa, A., Pihkola, H., Deviatkin, I., Tenhunen, A., Mannila, J., et al. (2020). Pyrolysis of plastic waste: Opportunities and challenges. *Journal of Analytical and Applied Pyrolysis, 152*, Article 104804.
Remadnia, A., Dheilly, R., Laidoudi, B., & Quéneudec, M. (2009). Use of animal proteins as foaming agent in cementitious concrete composites manufactured with recycled PET aggregates. *Construction and Building Materials, 23*(10), 3118–3123.
Richardson, A., & Landless, S. (2009). Synthetic fibres and steel fibres in concrete with regard to bond strength and toughness. *Northumbria Working Paper Series: Interdisciplinary Studies in the Built and Virtual Environment, 2*(2), 128–140.
Sabaa, B., & Ravindrarajah, R. S. (1997). Engineering properties of lightweight concrete containing crushed expanded polystyrene waste. *Paper presented at the Proceedings of the symposium MM: advances in materials for cementitious composites*. Boston, MA, USA.
Saikia, N., & Brito, Jd. (2013). Waste polyethylene terephthalate as an aggregate in concrete. *Materials Research, 16*, 341–350.
Saikia, N., & De Brito, J. (2014). Mechanical properties and abrasion behaviour of concrete containing shredded PET bottle waste as a partial substitution of natural aggregate. *Construction and Building Materials, 52*, 236–244.
Shamsuyeva, M., & Endres, H.-J. (2021). Plastics in the context of the circular economy and sustainable plastics recycling: Comprehensive review on research development, standardization and market. *Composites Part c: Open Access, 6*, Article 100168.
Sharma, R., & Bansal, P. P. (2016). Use of different forms of waste plastic in concrete–a review. *Journal of Cleaner Production, 112*, 473–482.
Sigvardsen, N. M. (2016). Use of discarded fishing nets as near surface mounted reinforcement for prolonging lifetime of existing structures. Paper presented at the Grøn dyst 2016.
Silva, D., Betioli, A., Gleize, P., Roman, H., Gomez, L., & Ribeiro, J. (2005). Degradation of recycled PET fibers in Portland cement-based materials. *Cement and Concrete Research, 35*(9), 1741–1746.
Silva, E., Coelho, J., & Bordado, J. (2013). Strength improvement of mortar composites reinforced with newly hybrid-blended fibres: Influence of fibres geometry and morphology. *Construction and Building Materials, 40*, 473–480.
Spadea, S., Farina, I., Carrafiello, A., & Fraternali, F. (2015). Recycled nylon fibers as cement mortar reinforcement. *Construction and Building Materials, 80*, 200–209.
Tota-Maharaj, K., Adeleke, B. O., & Nounu, G. (2022). Effects of waste plastics as partial fine-aggregate replacement for reinforced low-carbon concrete pavements. *International Journal of Sustainable Engineering, 15*(1), 192–207.
Vairagade, V., Kene, K., & Deshpande, N. (2012). Investigation of compressive and tensile behavior of fibrillated polypropylene fibers reinforced concrete. *International Journal of Engineering Research and Applications, 2*(3), 1111–1115.

Verma, R., Vinoda, K., Papireddy, M., & Gowda, A. (2016). Toxic pollutants from plastic waste-a review. *Procedia Environmental Sciences, 35*, 701–708.
Veropalumbo, R., Russo, F., Viscione, N., Biancardo, S. A., & Oreto, C. (2021). Investigating the rheological properties of hot bituminous mastics made up using plastic waste materials as filler. *Construction and Building Materials, 270*, Article 121394.
Wang, Y., Zureick, A.-H., Cho, B.-S., & Scott, D. (1994). Properties of fibre reinforced concrete using recycled fibres from carpet industrial waste. *Journal of Materials Science, 29*, 4191–4199.
Yoo, D.-Y., Park, J.-J., Kim, S.-W., & Yoon, Y.-S. (2013). Early age setting, shrinkage and tensile characteristics of ultra high performance fiber reinforced concrete. *Construction and Building Materials, 41*, 427–438.
You, L., Long, Z., You, Z., Ge, D., Yang, X., Xu, F., et al. (2022). Review of recycling waste plastics in asphalt paving materials. *Journal of Traffic and Transportation Engineering (English Edition), 9*(5), 742–764.
Yu, K., Li, L., Yu, J., Wang, Y., Ye, J., & Xu, Q. (2018). Direct tensile properties of engineered cementitious composites: A review. *Construction and Building Materials, 165*, 346–362.
Zulkernain, N. H., Gani, P., Chuan, N. C., & Uvarajan, T. (2021). Utilisation of plastic waste as aggregate in construction materials: A review. *Construction and Building Materials, 296*, Article 123669.

Plastic Pollution Disrupting the Aquatic Ecosystem

Yaksh S. Surela, Mital K. Lodariya, Debleena Bhattacharya, and K. R. Abhilash

Abstract Plastic pollution has created havoc for the ecosystem in recent years. With the advancement of technology, there is an increase in plastic pollution and it has contaminated all the ecological parameters. Plastic pollution has become an acute threat to aquatic ecosystems worldwide, unsettling marine and freshwater environments with far-reaching consequences. It is a complex, multi-layered issue that impacts all levels of the food web, from tiny plankton to large marine mammals, and affects biodiversity, ecosystem services, and even human health. Marine animals, like fish, seabirds, turtles, and mammals, frequently ingest plastic particles assuming it as food. When ingested, plastic can block digestive tracts, reduce appetite, and ultimately lead to malnutrition, starvation, or death. Entanglement in larger plastic items like fishing nets, bags, and rings often results in physical injuries, impaired movement, and drowning. Plastics release toxic chemicals, including Bisphenol A (BPA) and phthalates, which leach out into digestive system of the animals where in these toxins accumulate in their bodies, disrupting hormonal systems and affecting reproduction and development. Through this review, the multifaceted problem of plastic contamination is addressed. This study aims to review the plastic pollution, mitigation, and remediation of the damaged areas. Furthermore, after observation, the inference resides in the current plastic applications, benefits, adverse effect, and risk assessment, we augmented the inculcation of technological mitigation ways to curb this pollution from its roots. Through this study various pathways were also determined toward sustainable use of plastics and future research ways to solve this burgeoning problem.

Keywords Microplastic · Single-use plastic · Aquatic ecosystem · Climate change · Non-biodegradable waste

Y. S. Surela · M. K. Lodariya · D. Bhattacharya (✉)
Department of Chemical Engineering, Marwadi University, Rajkot, Gujarat, India
e-mail: debleena.b3@gmail.com

K. R. Abhilash
Conservation of Coastal and Marine Resources Division, National Centre for Sustainable Coastal Management, Ministry of Environment, Forest and Climate Change, Government of India, Chennai, Tamil Nadu, India

P. Pathak et al. (eds.), *Plastic Footprint*, https://doi.org/10.1007/978-981-96-6362-0_4

1 Introduction

Plastic was a boon to mankind when it was initially introduced to the market and rapidly became part of our daily requirements. The durability, resistance to decomposition, easy availability, and less cost added to its demand. With the advent of plastics, so many avenues opened for the industries and the major industry that benefitted is the packaging industry. Paper use was reduced, and plastic consumption soared. Plastic has reshaped industries and daily life, revolutionizing packaging, manufacturing, and consumer goods with its versatility, affordability, and durability. However, plastic's resilience has turned it into an enduring environmental hazard. Since the mid-twentieth century, plastic production has soared, reaching over 400 million tons per year in 2022, and over half of this output is single-use plastic, designed to be discarded after a short lifespan.

The mismanagement of plastic waste has become a global crisis, particularly in aquatic ecosystems. By 2024, approximately 8 million tons of plastic will enter the oceans annually. This vast influx affects marine biodiversity, disrupts ecosystems, and posed threats to human health. Plastic pollution contaminates coastal areas, open oceans, and even the most remote habitats, impacting marine species across the food chain—from microscopic plankton to apex predators. This chapter examines the history of plastic, the exponential growth of plastic pollution, and its devastating effects on the aquatic ecosystem. We also review recent research (2022–2024), analyze national and international responses, and propose future directions for combating this growing issue, including innovations like algae-based plastics.

1.1 History of Plastic and the Rise of Plastic Pollution

1.1.1 The Birth of Plastic

The story of plastic began in the nineteenth century as a response to the growing demand for natural materials such as ivory and tortoiseshell. Alexander Parkes, a British inventor, introduced Parke sine in 1862 as the first synthetic plastic made from organic cellulose. Although Parke sine did not achieve widespread use, it sets the foundation for the plastic industry. The breakthrough arrived in 1907 when Leo Baekeland, a Belgian chemist, created Bakelite, the first fully synthetic plastic derived from fossil fuels. Unlike previous plastics, Bakelite was entirely man-made and became widely adopted for its non-conductive and heat-resistant properties, making it ideal for electrical insulators and household products.

1.1.2 The Post-War Boom and Global Expansion

Following World War II, the production and consumption of plastic surged. With its adaptability and low cost, plastic became the alternative for traditional materials in everything from packaging to automotive manufacturing. During this era, the invention of single-use plastics transformed global markets, creating a "throwaway culture" where items like plastic bags, straws, and food containers were discarded after one use. By the 1970s, plastic had become ubiquitous, and single-use items were deeply ingrained in consumer habits. This disposable culture significantly contributed to the accumulation of plastic waste worldwide. As plastic production continued to climb, waste management and recycling systems struggled to keep pace, leaving much of this waste accumulated in natural environments, particularly oceans.

1.1.3 Plastic Pollution's Acceleration and Scale

The widespread use of plastic has led to persistent and pervasive pollution in the environment. Most plastics are non-biodegradable and can persist for hundreds of years, breaking down only into smaller particles, known as microplastics, rather than decomposing entirely. Microplastics, particles smaller than 5 mm, have become nearly impossible to remove from the environment and have infiltrated oceans, rivers, and anthropogenic food chains. Plastic waste enters oceans from land- and sea-based sources, including coastal littering, inadequate waste management, and lost fishing gear. The infamous Great Pacific Garbage Patch—one of several plastic gyres—contains around 1.8 trillion pieces of plastic and covers an area three times the size of France (LeBreton et al., 2018). These massive accumulations highlight the staggering scale of plastic pollution in the ocean and have reached unprecedented levels and continue to rise at an alarming rate. Without immediate action, the situation will only become more dangerous in the years ahead. Plastic waste is already harming numerous aquatic species and may also pose risks to human health in ways that remain largely unclear (Villarrubia-Gómez et al., 2018). The first scientific articles addressing plastic pollution in marine environments were published in 1969, highlighting the presence of plastics in the bodies of seabirds and other marine animals (Sighicelli et al., 2018).

2 Plastic Pollution and Its Disruption of the Aquatic Ecosystem

2.1 *Marine Wildlife Impacts*

Plastic pollution poses a direct threat to marine organisms across all trophic levels as shown in Fig. 1. Microplastics, for example, are often mistaken for food by small marine animals such as plankton, which form the foundation of the marine food web. Ingesting plastic particles can block digestive tracts, leading to malnutrition, reduced growth rates, and even death in some species (Rochman et al., 2013). This issue is not limited to plankton, as fish, seabirds, and marine mammals are also affected. Recent studies from 2022 to 2024 indicate that the ingestion of plastics can impair reproductive health, alter hormone levels, and increase stress in various marine species.

Plastic debris often causes harm to larger marine animals, such as turtles, seals, and dolphins. Plastic bags and balloons, for instance, are mistaken for jellyfish by turtles and can block their intestines, causing fatal obstructions. Additionally, entanglement in discarded fishing gear, known as "ghost nets," restricts movement, causing injury, starvation, and even death for species that become ensnared. A 2023 study reported that over 800 marine species are now affected by plastic pollution, including many endangered species.

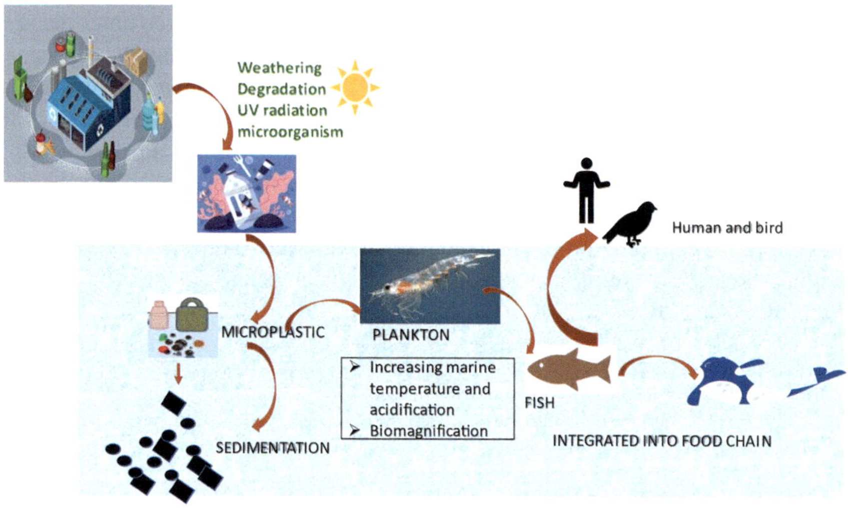

Fig. 1 Impact of plastic pollution on aquatic ecosystem

2.2 Ecosystem and Habitat Disruption

Plastic pollution not only harms individual organisms but also disrupts entire ecosystems. Coral reefs, known for their biodiversity, are particularly vulnerable. Plastic debris, when in contact with corals, causes physical abrasion, provides surfaces for pathogen colonization, and introduces diseases. A 2018 study published in science found that plastic debris increases the likelihood of coral diseases by 20-fold, putting already vulnerable coral reefs at greater risk of degradation. Seagrass beds and mangroves—vital habitats for many marine species—are similarly threatened. Plastics can smother these plants, disrupting nutrient cycling and reducing their ability to support marine biodiversity. Microplastics have even been found within sediments of coastal habitats worldwide, disrupting sediment structure and chemical processes (Law & Thompson, 2014).

2.3 Water Pollution by Plastics

Water pollution is driven by various factors, including untreated industrial waste, municipal waste, and the leaching of heavy metals from soil into groundwater. Both oceans and many land-based water systems are primarily polluted due to various human activities. Industrialization has led to the production of materials that have become pollutants due to their excessive production, widespread use, and inability to degrade. Plastics, particularly, is a leading contributor to global water pollution. This chapter focuses on the impact of plastics on water quality, specifically how they contaminate both marine and freshwater sources. It also explores potential solutions and measures to address and mitigate water pollution.

Plastics can be classified into two main categories based on their thermal properties: thermoplastics and thermosets. Thermoset plastics harden or set irreversibly when heated to a high temperature, making them known for their strength and durability. As a result, they are primarily used in the automotive and construction industries, as well as in applications such as adhesives, inks, and coatings. In contrast, thermoplastic polymers have weaker molecular bonds, which allow them to be softened and reshaped when heated. As a result, thermoplastics soften when heated by a flame and return to their original shape once cooled to room temperature. This property allows them to be easily molded and shaped into a variety of products, such as carpet fibers, credit cards, milk jugs, and floor coverings (Halden, 2010). Plastic fragments also vary in size, which leads to their classification into three categories: macro-fragments (greater than 5 mm), micro-fragments (ranging from 1 μm to 5 mm), and nano-fragments (less than 2 mm).

2.4 Macroplastics

Macroplastics refer to larger pieces of plastic, typically larger than 5 mm, that are commonly found in marine environments. According to Andrady (2011), the duration for which macroplastics remain in the ocean can vary significantly. For example, a plastic bag can last up to 50 years, a polyethylene bottle can persist for 450 years, cigarette butts can remain for 1 to 5 years, polystyrene cups for 50 years, plastic six-pack rings for 400 years, and fishing lines for up to 600 years.

2.5 Microplastics

Microplastics, like most synthetic plastics, are solid particles that can vary in shape, either organized or random, with sizes ranging from 1 μm to 5 mm. These particles are insoluble in water (Frias & Nash, 2019). Sharma and Chatterjee (2017) classified microplastics into two categories based on their origin: primary and secondary microplastics. Primary microplastics are small from the outset, whereas secondary microplastics are the result of larger plastic items breaking down into smaller fragments. The Joint Group of Experts on the Scientific Aspects of Marine Environmental Protection defines microplastics as "plastic particles less than 5 mm in diameter, including those in the nanoscale range (i.e., 1 nm)" (Beaumont et al., 2019).

According to Ogunola et al. (2018) and Andrady (2011), marine microplastic pollution originates from both primary and secondary sources. Primary microplastics come from everyday products such as toothpaste, cosmetics, detergents, shampoos, shower gels, and laundry fibers. Studies suggest that about 90% of microplastics are trapped in wastewater treatment plants (Law, 2017; Rhodes et al., 2018). Microplastics have spread rapidly across various ecosystems, from terrestrial habitats to the deep ocean, and have even been detected in polar regions.

Numerous marine species are at risk of becoming entangled in microplastics due to their small size, leading to physical injuries, starvation, and suffocation. This poses significant environmental concerns and threatens biodiversity. Secondary microplastics, on the other hand, form when larger plastic items degrade into smaller particles due to exposure to solar radiation and other environmental factors.

These augments the need for more research to both quantify plastic pollution and evaluate its economic, ecological, biological, and social consequences, as well as to identify potential solutions. Urban beaches and coastal waters are frequently covered in plastic debris, and microplastics have been discovered in nearly every habitat, from surface waters to beaches, the deep seafloor, and the water column. The environmental effects of plastic pollution include habitat degradation, health risks, and negative impacts on aquatic organisms (Bråte et al., 2017).

Plastic enters the oceans through a variety of pathways from multiple sources, yet the full range of origins contributing to ocean pollution remains unclear. Currently, it is not possible to provide reliable quantitative comparisons of plastic input levels, sources, or sectors responsible, revealing a significant gap in our understanding (UNEP, 2016). On land, plastics typically enter the ocean via shorelines, wastewater pipelines, or rivers. Additionally, waste management practices and certain weather events, such as storms and strong winds, can result in large amounts of plastic being swept into the ocean, posing significant challenges to address.

At sea, ocean-based sources of plastic pollution often arise from accidental losses, shipping activities—including cargo ships, oil tankers, and other vessels—and the deliberate disposal of plastic waste during maritime journeys (Bråte et al., 2017).

Land-based sources of macroplastics include food packaging, tourism, household products, construction materials, and drink containers. It is estimated that approximately 4.8 to 13 million metric tons of plastic enter the ocean each year from land-based sources. Marine-based sources of macroplastics primarily stem from the shipping industry and fisheries. Additionally, aquaculture infrastructure, which is largely made of plastic, can contribute to plastic pollution if the structures are poorly managed or damaged by environmental conditions (Haward, 2018).

2.6 Consequences of Plastic Pollution

Plastic enters the oceans through various routes from multiple sources, but the complete range of origins contributing to ocean pollution remains poorly understood. At present, it is impossible to make reliable quantitative comparisons between plastic input levels, sources, or sectors responsible, highlighting a significant knowledge gap (UNEP, 2016). On land, plastics typically make their way into the ocean via shorelines, wastewater systems, or rivers. Moreover, waste management practices, along with certain weather events like storms and high winds, can cause large quantities of plastic to be carried into the sea, presenting complex challenges.

At sea, plastic pollution often originates from accidental losses, shipping operations—such as those involving cargo ships, oil tankers, and other vessels—and the intentional disposal of plastic waste during maritime journeys (Bråte et al., 2017).

Land-based sources of macroplastics include food packaging, tourism-related waste, household products, construction materials, and beverage containers. It is estimated that between 4.8 and 13 million metric tons of plastic enter the ocean each year from these land-based sources. Marine-based sources of macroplastics primarily come from the shipping and fishing industries. Additionally, plastic-based infrastructure used in aquaculture can contribute to ocean pollution, especially if it is poorly maintained or damaged by environmental factors (Haward, 2018).

Aquatic organisms, including molluscs, birds, fish, crustaceans, and mammals, often mistake microplastics for food or intentionally feed on them instead of their natural diet. This not only directly impacts the creatures consuming microplastics but also disrupts entire trophic levels within the food web. Studies have shown that marine animals such as whales, fish, seabirds, and turtles often ingest plastic particles in their environment, leading to detrimental effects on their physiology and, in some cases, fatal outcomes. Plastic consumption can occur both directly and indirectly.

For instance, polyethylene bags floating in the water resemble jellyfish and are often mistaken for natural prey by turtles. Similarly, when a mussel consumes plastic, it results in direct ingestion, but if a crab eats the mussel, the plastic moves into its gastrointestinal tract, leading to indirect consumption. It has been confirmed that only a small portion of the ingested plastic passes through the digestive tract, with the rest being retained or stored in the body, causing a range of health issues (Lamb et al., 2018).

Microplastics ingested by living organisms can interfere with their normal physiological functions. For example, the animal's ability to feed may be impaired, which in turn affects its growth and reproduction. Experiments conducted in controlled environments have shown that microplastic ingestion by various aquatic organisms can have severe health consequences. Oysters exposed to polystyrene microparticles have shown a significant reduction in energy uptake, offspring performance, and reproductive success. Similarly, copepods that ingest polystyrene exhibit decreased egg hatching success and diminished algal feeding, leading to negative effects on their survival. Polyvinyl chloride (PVC) microplastics have been shown to harm soil-dwelling worms by depleting their energy reserves and disrupting their normal physiological functions (Eerkes-Medrano et al., 2015). These findings clearly highlight the profound impact synthetic plastics have on organisms, potentially altering the functioning of marine ecosystems.

2.7 *Species Affected by Plastic Pollution*

Plastic pollution has put many species, both on land and in water, in grave danger, with particularly alarming effects on marine life. The reason marine organisms are so vulnerable is their tendency to mistake plastics for food or, in some cases, selectively feed on plastic particles present in their habitats. The sea ecosystem is particularly vulnerable to plastic pollution because nearly 80% of plastic waste generated on land ends up in water resources. Once in the water, plastics break down into smaller fragments and particles of varying sizes due to the movement of water currents and other marine factors. Numerous studies have examined the effects of plastics on aquatic organisms, focusing on how they impact normal biological functions, physiological activities, and the overall health of marine habitats. In this section, we will highlight several marine species to illustrate the devastating consequences faced by sea life and the ongoing struggle for survival between marine organisms and plastics in the oceans.

Sea turtles are frequently studied to assess the harmful effects of plastic pollution on marine life. Numerous autopsies and biological analyses have shown that plastic and tar are among the primary causes of stress and unnatural death in sea turtles, as they often ingest these materials while living in polluted waters, especially in oceans. Common plastic debris found in the digestive tracts of turtles includes plastic bags, polyethylene terephthalate (PET) bottle fragments, and plastic ropes. Studies have also revealed that sea turtles often mistake polyethylene bags for jellyfish, their natural prey. This misidentification leads to the ingestion of plastic, which ultimately contributes to their death (Sigler, 2014).

The International Union for Conservation of Nature (IUCN) has listed the leatherback sea turtle as critically endangered due to a significant decline in its population over the past two decades. In a study, autopsies were performed on 371 leatherback turtles collected from their natural habitats, revealing that 137 of them had plastic in their digestive systems. However, it was not definitively concluded that the ingested plastic was the sole cause of death, as other factors may have been involved. Studies have shown that 8.7% of turtles that died from plastic pollution had a plastic or polyethylene bag blocking their esophagus, preventing food from reaching their stomach, which ultimately led to their death (Vegter et al., 2014).

In addition to the immediate threats posed to sea life, there has been a noticeable decline in the reproductive success of these animals. Research on female turtles has shown that plastic can obstruct the passage of eggs. In one case, 14 different types of plastics were recovered from the cloaca of a female turtle, which ultimately hindered egg-laying (Sigler, 2014).

Seafood is considered a delicacy worldwide, with people consuming a variety of marine species such as lobsters, crabs, oysters, and fish. Approximately 19% of the global population relies on seafood to meet their animal protein needs (Beaumont et al., 2019). As apex predators in the food web, humans are at the top trophic level, which means they indirectly consume plastics through the seafood they eat—particularly shellfish and small fish, which are often consumed whole, including their digestive systems. In contrast, larger fish, which have more flesh, are typically eaten for their meat, making humans less directly exposed to plastics in these species.

Studies and experiments have shown that nano-plastics are more mobile in the tissues of living organisms compared to other forms of plastic. Their larger surface-to-volume ratio increases their concentration in tissues, raising concerns about potential health and reproductive issues if seafood contaminated with nano-plastics is consumed. As plastic nanoparticles are widely distributed in the marine environment, with unknown sources of origin, the growing presence of plastic pollution in aquatic ecosystems calls for greater awareness. Given these risks, humans need to be cautious about the potential health impacts of consuming seafood contaminated by plastics (Rhodes, 2018).

3 National and International Responses (2022–2024)

Various countries have introduced measures to curb plastic use and pollution, primarily by focusing on single-use plastics. In 2023, the European Union implemented a directive banning single-use plastic items like straws, cutlery, and plates. This move has significantly reduced marine litter in Europe, particularly along coastlines. India introduced a nationwide ban on certain single-use plastics in 2022, targeting the most commonly discarded items. Japan, a leading consumer of plastics in Asia, has focused on enhancing recycling rates, investing in waste-to-energy technologies, and promoting bioplastics. Meanwhile, African nations like Rwanda and Kenya have taken pioneering steps in banning plastic bags entirely, positioning themselves as leaders in plastic waste reduction (Desai, 2020). Plastic pollution is also a transboundary issue, making international cooperation essential. The Global Plastics Treaty, under negotiation and expected to be finalized by 2024, represents a landmark effort to address plastic pollution globally. This treaty will require participating countries to implement strategies for plastic reduction, improve waste management, and encourage innovation in biodegradable alternatives. If successful, it will mark a turning point in global plastic policy and set enforceable guidelines for plastic management. International organizations like the World Wildlife Fund (WWF) and Greenpeace have also raised public awareness, advocating for plastic-free alternatives and urging consumers to adopt sustainable lifestyles. These campaigns, alongside the efforts of global brands pledging to reduce plastic use, have helped to shift public attitudes and promote behavioral change on a large scale.

4 Treatment and Management of Plastic Waste

4.1 Current Methods of Plastic Waste Management

Traditional method for managing plastic waste, such as landfilling, incineration, and recycling, shows significant limitations. Landfills contribute to environmental contamination and leach harmful chemicals into the soil and groundwater. Incineration, while reducing waste volume, releases toxic gases, including dioxins, which have adverse health effects. Recycling rates remain low worldwide; only about 9% of all plastic produced is recycled, with the remainder entering landfills, incinerators, or natural environments. Even in regions with advanced recycling systems, only specific types of plastic are accepted, limiting overall effectiveness.

4.2 Innovative Solutions for Plastic Waste Treatment

As awareness of plastic pollution grows, researchers are exploring novel solutions. Biodegradable plastics, for example, offer promise as an alternative to traditional plastics. These materials are often made from renewable sources, like cornstarch or sugarcane, and are designed to decompose more rapidly than conventional plastics. However, they still require specific conditions to break down and may not fully address pollution in natural settings. Algae-based plastics are a cutting-edge innovation showing potential as a sustainable alternative. Algae, a fast-growing and renewable resource, can be processed into bioplastic that biodegrades without leaving microplastics. Researchers are also studying microbial degradation, in which certain bacteria and fungi can consume plastic waste, breaking it down more efficiently. Although still in development, these technologies could offer a scalable solution to plastic waste.

4.3 *Remediation for the Plastic Pollution*

The continued utilization of plastics can pose a directly harmful impact on the ecosystem and ecological environment. That can generate crises regarding plastic waste management and disposal. So many techniques like incineration, pyrolysis, recycling, and degradation are still emerging to reduce plastic pollution. However, these techniques are of high cost and unsustainable. Thus, environment-friendly, cost-effective, sustainable solutions have been required. Due to the increasing anthropogenic activities, urbanization, and industrialization a large amount of plastic waste have been generated which is landfilled, resulting in water and soil pollution. Plastic waste is converted into smaller pieces due to physical, chemical, and biological processes known as micro-nano-plastics (MNPs). They can physically harm aquatic and terrestrial life through ingestion. Major sources of micro- and nanoparticles are wastewater treatment plants from cosmetics, synthetic micro-fibers from laundry, and other household plastic waste. Zhen et al. (2024) have investigated the remediation and detection of microplastics through efficient interfacial solar evaporation for immaculate water production. The result indicates microplastic absorbance ratio has been increasing up to five times as compared to the previous study. Thus, our rationally designed ISEP holds promising potential to not only mitigate the existing water scarcity issue but also remediate MP's pollution in natural water environments. Smart and advanced technology, such as microbial and catalytic methods, has yielded exciting results. However, these methods cannot directly address MPs in a natural environment unless pre-recycling, given the low concentration of MPs. Some physical technologies have emerged as effective solutions for treating MPs. Microbubbles can be produced from MPs through passing sunlight using glass balls (Wang et al., 2022). One of the best methods for eliminating microplastics from tap water is to use a water filter, such as one made of ceramic, activated carbon, or reverse osmosis. Nanomaterial is also used for remediation techniques of plastics as they have excellent adsorption capacity of nano-plastics and microplastics from water. One promising approach is the use of microbial-based remediation. Researchers have isolated and identified microorganisms from agricultural soils that can degrade different types of plastics, including polyethylene polymers. The remediation of plastic pollution involves various techniques aimed at reducing, removing, or mitigating the harmful effects of plastics in the environment. These techniques vary depending on the type of plastic and the environmental setting—whether land, freshwater, or marine ecosystems (refer Table 1).

Table 1 The remediation techniques and plastic type

S. no	Remediation techniques	Plastic type	Benefits	References
1	Reactive compression molding	Sulfur polymers	1. Made soft and flexible nature of the rubber 2. Recycling of a 50-poly(S-rcanola) mat	Nicholas et al. (2020)
2	Biological degradation	Polyethylene terephthalate (PET)	Metabolizing PET and use it as a carbon source	Moog et al. (2019)
3	Chemical recycling	Polystyrene (PS)	Breaks down PS into monomers for reuse in new products	Rahimi and García (2017)
4	Pyrolysis	Mixed plastics	Converts plastics into fuels and other valuable chemicals	Zhang et al., (2020)
5	Enzymatic recycling	Polyurethane (PU)	It uses enzymes to depolymerize PU into its chemical components	Tournier et al. (2020)
6	Solvent-based recycling	Polyvinyl Chloride (PVC)	Dissolves PVC for separation and recovery of pure plastic	Deng et al. (2021)
7	Mechanical recycling	Polypropylene (PP)	Reduces plastic waste by remolding into new products	Ragaert et al. (2017)
8	Photocatalytic degradation	Polyethylene (PE)	Uses light to degrade PE into smaller molecules	Sajjadi et al. (2020)
9	Hydrothermal processing	Polyamide (PA)	Converts PA into valuable chemicals under high-pressure water conditions	Lopez et al. (2020)
10	Microwave-assisted degradation	Polymethyl Methacrylate (PMMA)	Accelerates degradation of PMMA using microwave energy	Thakur et al. (2023)
11	Supercritical fluid degradation	Polystyrene (PS)	Uses supercritical fluids to break down PS into reusable monomers	Yoshida et al. (2016)
12	Biodegradation by fungi	Polyethylene Terephthalate (PET)	Uses fungi to metabolize and break down PET	Urbanek et al. (2018)
13	Thermal cracking	Mixed plastics	Breaks down plastics into oil and gas through high-temperature treatment	Lomonaco et al. (2021)
14	Ionic liquid-assisted recycling	Polyethylene Terephthalate (PET)	Uses ionic liquids to dissolve and recover PET	Wei et al. (2020)
15	Catalytic cracking	Polypropylene (PP)	Converts PP into valuable hydrocarbons using catalysts	Panda et al. (2024)

(continued)

Table 1 (continued)

S. no	Remediation techniques	Plastic type	Benefits	References
16	Ultrasonic degradation	Polyvinyl Alcohol (PVA)	Uses ultrasonic waves to break down PVA into smaller molecules	He et al. (2021)
17	High-energy radiation degradation	Polyethylene (PE)	Uses high-energy radiation to degrade PE	Singh et al. (2018)
18	Composting with additives	Polylactic Acid (PLA)	Enhances composting process of PLA with specific additives	Tsuji and Suzuyoshi (2020)
19	Fluidized bed pyrolysis	Mixed plastics	Converts plastics into fuels and chemicals in a fluidized bed reactor	Sharuddin et al. (2016)
20	Gasification	Mixed plastics	Converts plastics into syngas for energy production	Chen et al. (2020)

5 Conclusions

Plastic pollution poses a significant environmental threat to the planet. It is not only a growing concern for humans but also for all life forms within the biosphere. The rising production and continuous use of various plastic products have led to a worsening situation. Plastic pollution is escalating at alarming levels, threatening not only our land but also our water resources, including rivers, lakes, and oceans. Addressing this issue requires urgent and concerted efforts. To protect our water bodies from plastic pollution, we need a comprehensive strategy that focuses on three key solution areas. The advancements in these fields can help us develop bioplastics that are biodegradable and eco-friendly, providing an alternative to synthetic plastics. By replacing conventional plastics with biodegradable options, we can reduce their production and, consequently, their environmental impact on both land and water. Further, the most critical domain is the formulation and enforcement of strong government policies, regulations, laws, and international agreements aimed at environmental conservation. National and international bodies must collaborate to create and implement comprehensive frameworks to combat plastic pollution.

In summary, to safeguard our water resources from plastic pollution, we must focus on all three solution areas: science and technology, education and community awareness, and robust government policies at both the national and international levels.

References

Andrady, A. L. (2011). Microplastics in the marine environment. *Marine Pollution Bulletin, 62*(8), 1596–1605.

Beaumont, N. J., Aanesen, M., Austen, M. C., Börger, T., Clark, J. R., Cole, M., & Wyles, K. J. (2019). Global ecological, social and economic impacts of marine plastic. *Marine Pollution Bulletin, 142*, 189–195.

Bråte, I. L. N., Eidsvoll, D. P., Steindal, C. C., & Thomas, K. V. (2016). Plastic ingestion by Atlantic cod (Gadus morhua) from the Norwegian coast. *Marine Pollution Bulletin, 112*(1–2), 105–110.

Bråte, I. L. N., Huwer, B., Thomas, K. V., Eidsvoll, D. P., Halsband, C., Almroth, B. C., & Lusher, A. (2017). *Micro-and macro-plastics in marine species from Nordic waters*. Nordic Council of Ministers.

Chen, S., Liu, Z., Jiang, S., & Hou, H. (2020). Carbonization: A feasible route for reutilization of plastic wastes. *Science of the Total Environment, 710*, 136250.

Deng, H., Li, R., Yan, B., Li, B., Chen, Q., Hu, H., ... & Shi, H. (2021). PAEs and PBDEs in plastic fragments and wetland sediments in Yangtze estuary. *Journal of Hazardous Materials, 409*, 124937.

Desai, B. H. (2020). 14. United nations environment programme (UNEP). *Yearbook of International Environmental Law, 31*(1), 319–325.

Eerkes-Medrano, D., Thompson, R. C., & Aldridge, D. C. (2015). Microplastics in freshwater systems: A review of the emerging threats, identification of knowledge gaps and prioritisation of research needs. *Water Research, 75*, 63–82.

Frias, J. P., & Nash, R. (2019). Microplastics: Finding a consensus on the definition. *Marine Pollution Bulletin, 138*, 145–147.

Halden, R. U. (2010). Plastics and health risks. *Annual Review of Public Health, 31*(1), 179–194.

Haward, M. (2018). Plastic pollution of the world's seas and oceans as a contemporary challenge in ocean governance. *Nature Communications, 9*(1), 667.

He, D., Zhang, Y., & Gao, W. (2021). Micro (nano) plastic contaminations from soils to plants: Human food risks. *Current Opinion in Food Science, 41*, 116–121.

Lamb, J. B., Willis, B. L., Fiorenza, E. A., Couch, C. S., Howard, R., Rader, D. N., & Harvell, C. D. (2018). Plastic waste associated with disease on coral reefs. *Science, 359*(6374), 460–462.

Law, K. L. (2017). Plastics in the marine environment. *Annual Review of Marine Science, 9*(1), 205–229.

Law, K. L., & Thompson, R. C. (2014). Microplastics in the seas. *Science, 345*(6193), 144–145.

Lebreton, L., Slat, B., Ferrari, F., Sainte-Rose, B., Aitken, J., Marthouse, R., Hajbane, S., Cunsolo, S., Schwarz, A., Levivier, A., & Noble, K. (2018). Evidence that the great pacific garbage patch is rapidly accumulating plastic. *Scientific Reports, 8*(1), 1–15.

Lomonaco, V., Pellegrini, L., Cossu, A., Carta, A., Graffieti, G., Hayes, T. L., ... & Maltoni, D. (2020). *CVP 2020 continual learning in computer vision competition: approaches, results, current challenges and future directions*. Artificial Intelligence, Advance online publication.

Lopez, Y. M., Paes, J. B., Gustave, D., Gonçalves, F. G., Méndez, F. C., & Nantet, A. C. T. (2020). Production of wood-plastic composites using cedrela odorata sawdust waste and recycled thermoplastics mixture from post-consumer products-A sustainable approach for cleaner production in Cuba. *Journal of Cleaner Production, 244*, 118723.

Moog, D., Schmitt, J., Senger, J., Zarzycki, J., Rexer, K. H., Linne, U., & Maier, U. G. (2019). Using a marine microalga as a chassis for polyethylene terephthalate (PET) degradation. *Microbial Cell Factories, 18*, 1–15.

Nicholas, J. M., & Steyn, H. (2020). *Project management for engineering, business and technology*. Routledge.

Ogunola, O. S., Onada, O. A., & Falaye, A. E. (2018). Mitigation measures to avert the impacts of plastics and microplastics in the marine environment (a review). *Environmental Science and Pollution Research, 25*, 9293–9310.

Panda, A. K., Rout, S. K., & Das, A. K. (2024). Optimization of diesel engine performance and emission using waste plastic pyrolytic oil by ANN and its thermo-economic assessment. *Environmental Science and Pollution Research, 31*(27), 38893–38907.
Ragaert, K., Delva, L., & Van Geem, K. (2017). Mechanical and chemical recycling of solid plastic waste. *Waste Management, 69*, 24–58.
Rahimi, A., & García, J. M. (2017). Chemical recycling of waste plastics for new materials production. *Nature Reviews Chemistry, 1*(6), 0046. https://doi.org/10.1038/s41570-017-0046.
Rhodes, C. J. (2018). Plastic pollution and potential solutions. *Science Progress, 101*(3), 207–260.
Rochman, C. M., Hoh, E., Kurobe, T., & Teh, S. J. (2013). Ingested plastic transfers hazardous chemicals to fish and induces hepatic stress. *Scientific Reports, 3*(1), 1–7.
Sajjadi, M., Baran, N. Y., Baran, T., Nasrollahzadeh, M., Tahsili, M. R., & Shokouhimehr, M. (2020). Palladium nanoparticles stabilized on a novel Schiff base modified Unye bentonite: Highly stable, reusable and efficient nanocatalyst for treating wastewater contaminants and inactivating pathogenic microbes. *Separation and Purification Technology, 237*, 116383.
Sharma, S., & Chatterjee, S. (2017). Microplastic pollution, a threat to marine ecosystem and human health: a short review. *Environmental Science and Pollution Research, 24*, 21530–21547.
Sharuddin, S. D. A., Abnisa, F., Daud, W. M. A. W., & Aroua, M. K. (2016). A review on pyrolysis of plastic wastes. *Energy Conversion and Management, 115*, 308–326.
Sighicelli, M., Pietrelli, L., Lecce, F., Iannilli, V., Falconieri, M., Coscia, L., & Zampetti, G. (2018). Microplastic pollution in the surface waters of Italian Subalpine Lakes. *Environmental Pollution, 236*, 645–651.
Sigler, M. (2014). The effects of plastic pollution on aquatic wildlife: Current situations and future solutions. *Water, Air, & Soil Pollution, 225*, 1–9.
Singh, N., et al. (2018). Recycling of plastic solid waste: A state of art review and future applications. *Composites Part b: Engineering, 115*, 409–422. https://doi.org/10.1016/j.compositesb.2016.09.013.
Thakur, B., Singh, J., Singh, J., Angmo, D., & Vig, A. P. (2023). Biodegradation of different types of microplastics: Molecular mechanism and degradation efficiency. *Science of the Total Environment, 877*, 162912.
Tournier, V., Topham, C. M., Gilles, A., David, B., Folgoas, C., Moya-Leclair, E., & Marty, A. (2020). An engineered PET depolymerase to break down and recycle plastic bottles. *Nature, 580*(7802), 216–219.
Tsuji, H., & Suzuyoshi, K. (2002). Environmental degradation of biodegradable polyesters 1. Poly (ε-caprolactone), poly [(R)-3-hydroxybutyrate], and poly (L-lactide) films in controlled static seawater. *Polymer Degradation and Stability, 75*(2), 347–355.
Unep, A., & Assessment, I. R. R. (2016). *The rise of environmental crime*. Nairobi: UNEP.
Urbanek, A. K., et al. (2018). Biodegradation of plastics by fungi: A review. *Biodegradation, 29*(4), 415–426. https://doi.org/10.1007/s10532-018-9826-4.
Vegter, A. C., Barletta, M., Beck, C., Borrero, J., Burton, H., Campbell, M. L., & Hamann, M. (2014). Global research priorities to mitigate plastic pollution impacts on marine wildlife. *Endangered Species Research, 25*(3), 225–247.
Villarrubia-Gómez, P., Cornell, S. E., & Fabres, J. (2018). Marine plastic pollution as a planetary boundary threat–The drifting piece in the sustainability puzzle. *Marine Policy, 96*, 213–220.
Wang, K., Chen, W., Tian, J., Niu, F., Xing, Y., Wu, Y., ... & Xu, L. (2022). Accumulation of microplastics in greenhouse soil after long-term plastic film mulching in Beijing, China. *Science of the Total Environment, 828*, 154544.
Wei, R., Tiso, T., Bertling, J., O'Connor, K., Blank, L. M., & Bornscheuer, U. T. (2020). Possibilities and limitations of biotechnological plastic degradation and recycling. *Nature Catalysis, 3*(11), 867–871.

Yoshida, S., Hiraga, K., Takehana, T., Taniguchi, I., Yamaji, H., Maeda, Y., & Oda, K. (2016). A bacterium that degrades and assimilates poly (ethylene terephthalate). *Science, 351*(6278), 1196–1199.

Zhen, Y., Sun, T. T., Wei, J. B., Zheng, Z. Y., & Chen, H. (2024). Radial oscillations of strange quark stars admixed with dark matter. *Symmetry, 16*(7), 807.

Development of a Standardized Plastic Footprint Measurement Tool for Combating Marine Plastic Pollution

M. S. S. R. Tejaswini and Puspita Dalui

Abstract Globally, marine ecosystem and its associated aquatic life have been seriously threatened due to the increasing disposing of plastic wastes into the marine environment. Although due to the versatility, malleability and cost-effectiveness have increased its production and consumption, it is eventually discarded as waste. The upsurging of micro–nano plastics into the environment demonstrates the systemic failure of the take-make-dispose consumption/linear model and draws the attention towards the need for a paradigm shift towards circular material flow models. Furthermore, there are various methodologies reported by the various companies, organization's, institutions and governments in order to tackle the plastic pollution, there is no well-established method to measure the extent of marine plastics that have been unscientifically disposed into the oceans. Accordingly, this chapter emphasizes on impact of plastics on the environment, existing and emerging plastic footprint methodologies. The environmental impact assessment and circular economy concepts should be used synergistically implemented to identify the best scenarios that would ultimately help in identification and quantification of marine plastics. Additionally, it also highlights the role of stakeholders that will help the policy makers and decision makers in understanding the objectives associated with different footprint methodologies.

Keywords Plastic impacts · Plastic footprint · Pollution · Metrics

1 Introduction

Plastics are anthropogenic materials which comprise diverse organic polymers with the ability to be given shapes of one's choice. One of its properties being, plasticity, furnished by the additives used, make it an excellent source for the manufacture of

M. S. S. R. Tejaswini (✉)
Department of Chemistry, School of Applied Sciences and Humanities, Vignan's Foundation for Science, Technology and Research, Guntur, Andhra Pradesh, India
e-mail: tejaswinialc@gmail.com

P. Dalui
Department of Zoology, Nizam College, Osmania University, Hyderabad, Telangana, India

P. Pathak et al. (eds.), *Plastic Footprint*, https://doi.org/10.1007/978-981-96-6362-0_5

different products (Cássio et al., 2022; Ahmed et al., 2021). Due to its malleability, durability, low cost, and easily available nature, its consumption and production has been increased in recent years. Products manufactured include cell phones, food packaging, blood bags, shopping bags, chairs, vehicles, and toys using polymers that accurately depict their advantageous qualities. Borrelle et al. (2020) reported that approximately 4,600 million metric tonnes of discarded plastic wastes have entered the both terrestrial and aquatic environments, that eventually upsurges the formation of secondary pollutants-micro–nano plastics (Gewert et al., 2015). The continuous leakage of plastic materials inside the oceans demonstrates the lack of effective waste management in developing countries. The non-biodegradable nature of plastics further adds to the challenge in managing the wastes in a sustainable manner (Chamas et al., 2020; Pujara et al., 2019).

It is estimated that by 2050, the cumulus of greenhouse gas release from plastic might outstretch above fifty-six gigatons (GT) of the enduring carbon budget (NITI Aayog, 2022). According to a study, polyethylene is predominantly utilized in India (Emami et al., 2024). The saline water bodies are so extensive that they extend over 71% of the planet. This ecosystem has plentiful flora and fauna that extend from elementary to pioneer life forms. It's investigated that 5 tonnes plastic remnants stay afloat on the oceans due to inappropriate scrapping of plastic waste (Thushari & Senevirathna, 2020). At present, pollution caused by plastics is a crucial issue in many marine bodies of the sphere (Govani et al., 2020; Tejaswini et al., 2022a). Moreover, the geographic location affects the extent of leakage from different types of polymers' leakage. In general, different grades of plastics are manufactured with the addition of potentially hazardous chemical additives and chemicals of high concern (CoHC), that are eventually leaked and bring about severe contamination into the environment.

Businesses, industries, hospitals, and institutions are increasingly seeking interest in evaluating the impacts of different plastics products that are being utilized in these organizations (Linder et al., 2017). However, to evaluate the significant impacts during each stage of life cycle of the plastics and to make informed decisions, adoption of accurate and reliable metrics becomes utmost priority. The plastic footprint is similar to the concept of a "carbon footprint," but specifically measures the environmental impact of plastic consumption and waste. This includes the evaluation of emissions during manufacturing, production, transportation, distribution and disposal stages of various plastic products, as well as the pollution caused by plastic waste, especially when it ends up in oceans, rivers, and landfills (Saidani et al., 2017). The term highlights the cumulative effect of plastic on the environment, encouraging individuals and organizations to reduce plastic usage, increase recycling, and promote sustainable alternatives (Rossi & Heine, 2014). A wide variety of footprint methodologies have been constructed in the former twenty years to notify the public, companies and policymakers about the expanse of utilization and manufacturing of plastic bottles influencing the environment as can be seen from Fig. 1.

The idea of footprint may cover three elements resulting in the following sets of metrics (Boucher et al., 2019).

- The amount of plastic used in an arrangement.

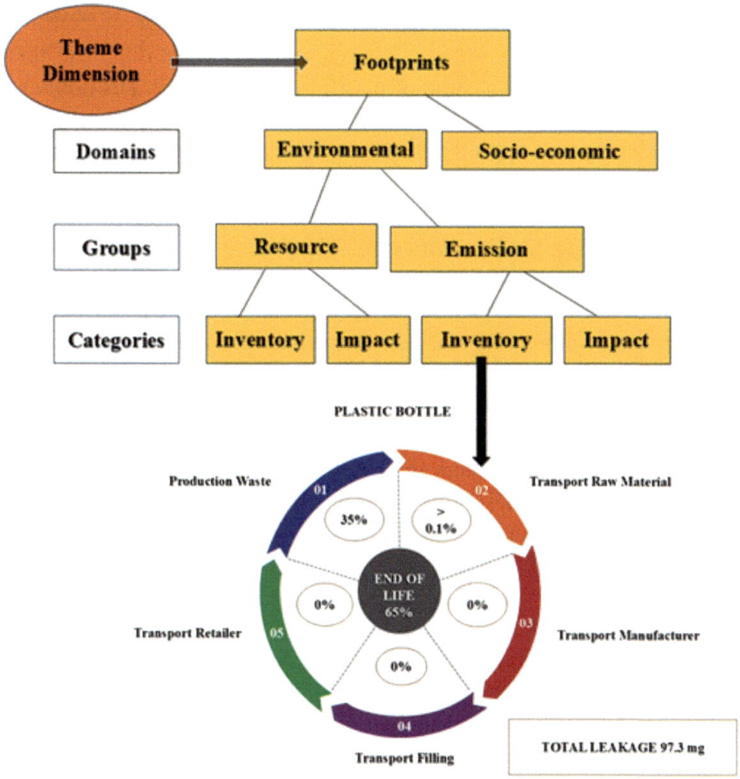

Fig. 1 Depiction of different types of environmental footprints obtained during the processing and utilization of one plastic bottle (Boucher et al., 2019; CPA, 2014)

- The aggregates of plastic being discharged into the environment in the course of making, transfer, usage or expiry of plastic items.
- The impression directly or indirectly created by the pollutants discharged on the health of human or the environment.

Although there are several existing methods that address certain aspects for aforementioned life cycle regions of interest, but they don't concentrate on the inherent risks associated with the chemicals that are used to make polymers and found in plastic products. The Plastics scorecard provides a method for assessing plastics' chemical footprint and a recommendation for safer substitutes. Globally, plastics production nearly requires 64 million pounds of chemicals, out of which 60% are chemicals of high concern (CoHCs), thereby increasing the chemical footprint. Chemical foot printing is the method for assessing the progress towards the shift and utilization of safer and greener chemicals away from hazardous chemicals. Accordingly, this chapter aims to identify the challenges in reducing the footprint, evaluate the different footprint methodologies, and understand the role of circularity indicators that will help in minimization of impacts.

2 Environmental Impacts Associated with Plastic Waste

Open dumping or direct combustion of plastics ultimately liberates noxious fumes such as dioxins, cyclic-chlorinated hydrocarbons besides greenhouse gases in our surroundings thereby causing climate change. Besides inappropriate handling of plastic makes them set foot in marine bodies including various other territories by many processes taking place in the domain. These, then stay in the habitat and unfold as a pollutant, microplastic (Tejaswini et al., 2022b).

A study reported those severe meteorological conditions like cyclone shifted these particles from terrestrial to aquatic habitat (especially marine bodies). Also those particles with density more than marine water submerged remaining particles stayed afloat (Dalui & Jyothi, 2021). Plastic particles of micro size associate with many unsafe-polluting substances like; PCB (Polychlorinated Biphenyls) and Cadmium via operations such as hydrogen-bonding, electrostatic interaction. Almost 45 K oceanic vertebrates are put through bioaccumulation of these toxic matter through processes like: engulfing microscopic and nanoscopic particles that ensure losing any extremity or an organ also organisms getting tangled in plastics. Oceanic plastic amasses simulates natural abode of oceanic organisms thereby allowing them to gather and colonize on it (Tejaswini et al., 2022a, 2022b). The various impacts of plastics in the environment can be seen from Fig. 2.

Immoderate loads of micro-sized plastic are known to reform the phytoplankton along with primary production. This can change the food web additionally affecting the functioning and interaction of the biotic and abiotic factors (Vivekanand et al., 2021). The exudation of substances from one hundred seventy-seven rivers, reach the mangrove biome of the Sundarbans thus allowing these synthetic polymers to settle as deposits also get collected in surface-water. Plastic particles, in large volumes get deposited in this delta due to many cyclonic episodes. Polycyclic Aromatic Hydrocarbons (PAHs) as well as Poly Chlorinated Biphenyls (PCBs) detected in

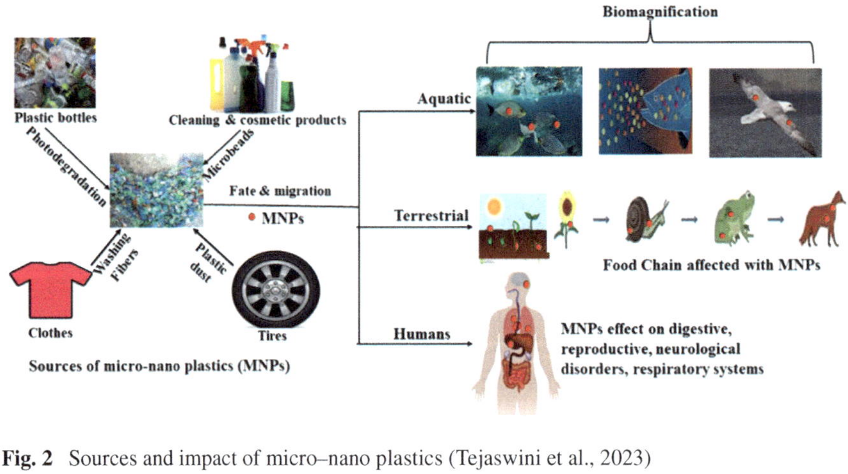

Fig. 2 Sources and impact of micro–nano plastics (Tejaswini et al., 2023)

the Sundarban delta effects the geochemistry of the mangrove forests along with its coastal ecology (Kumar et al., 2022). It is to be noted that eighty percent of oceanic remnant contains fifty percent fishing linked and thirty percent domestic or industrial-based plastics. It was reported that oceanic plastic amasses documented maximal mean estimate at Veraval (55.24 ± 16.52 kg/km^2, limits 0–162.5 kg/km^2) along with minimal mean estimate at Vishakhapatnam (2.11 ± 0.53 kg/km^2, limits 0.29–6.46 kg/km^2) (Kaladharan et al., 2020).

The result of plastic pollution on oceanic creatures discusses them being trapped in the oceanic amass; frequently in dropped or misplaced fishing apparatus. The trapped creatures are obstructed from locomotion, nourishment and respiration. The first report of an oceanic organism being trapped in an oceanic amass was a shark. These synthetic polymers are not assimilative hence, get collected in the organism's gut, thereby modifying fitness therefore, impacting propagation and existence. In turtles, ensnaring leads to epidermal septicity also, sever of limbs (Kühn et al., 2015). Plastic scrap is known to influence the coral reefs also. More than 1000 million remnants of plastic trap the corals, according to an inspection of one hundred fifty-nine coral reefs covering the Asia–Pacific region and this estimate is supposed to surge forty percent by 2025.

Plastic pollution develops disorders in corals, some of the maladies include white syndromes, skeletal eroding band, atramentous necrosis (Akhtar et al., 2022; Unice et al., 2018). The seagrass ecosystem serves as sinks to microplastics. These particles infiltrate into this ecosystem via coastal run-off as well as sewage let out. The microplastics adheres to the leaf blades of seagrasses, these then enter the bodies of plant-eating animals living in the oceanic habitat. Microplastics affect nutrient cycling by obstructing diazotroph dinitrogen fixation and limiting microbial processes along with decreasing root-nutrient uptake, thus, intervening with its microbiome (Ameen et al., 2024).

3 Plastic Footprint Methodologies

3.1 Circularity Indicators

According to the Unice et al. (2018), the methodology for determining the Material Circularity Index (MCI) will assess the paths and flow of material involved during its life cycle, besides taking into account the product's usefulness and lifespan. This will also include the quantity of recoverable, biodegradable, and decomposing materials, inaccessible waste, and reuse and recycling effectiveness of the process. This method considers a product to be totally linear if it is made entirely of virgin raw materials and is then disposed of in a landfill, meaning that MCI = 0. A product is deemed completely circular, or MCI = 1, if it uses no virgin raw materials and is collected for recycling its constituent parts with a 100% recycling efficiency (Saidani et al., 2017).

These indicators usually correlate between the entire economic worth of the product before recycling to the economic worth after recycling and recovering the product. The percentage that comes from used products determines the product's circularity, which can range from 0 to 1 (Linder et al., 2017; WEF, 2016). In this case, 1 indicates that every product gets recycled, producing an equivalent amount of value to the original. However, this indicator solely takes into account the recovered item's monetary worth; compared to the perspective of the environment, it has an extremely limited viewpoint and fails to evaluate any measure of authentic circularity. Using this indication to identify a product that is mechanically recovered could be deemed upcycling if recycling is thought to provide added value to the product with regard to its restoration (Verschoor et al., 2016).

Natural contamination of plastic emphasizes the need for transition for greater cyclic materials circulation and indicates that the take-make-dispose consumer paradigm has been undermined on a systemic level. Plastics require an immense material flow, beginning with natural gas and crude oil and culminating with chemicals and the finished product as can be seen from Table 1. In this context, increasing the circularity of plastic flows is crucial, but it is not the sole strategy to deal with the challenge. A widely recognized and comprehended approach to determining a plastic footprint might provide stakeholders and policymakers in conjunction the value chain associated with plastics the clarity that is required to take well-informed action to stop plastic leaks. Environmental footprints evaluate each of the emissions and pollutants resulting from a product or service throughout its life cycle, including direct and indirect sources (Unice et al., 2018). Plastic leakage is quantified when calculating a plastic footprint. The current investigation in the chapter, describes two approaches to conducting such dimensions: life cycle analysis (LCA) and Environmentally Extended Input Output Analysis (EEIOA). These approaches can provide an inventory of leakage, as well as an estimate of its impact and monetary value.

3.2 Plastic Footprint for Companies

There are different methodologies which require a systematic approach which will help different corporate and private organizations for measuring, assessing, quantifying and managing their environmental impacts related to consumption of plastics at each stage of their product manufacturing which has been summarized in Table 2. Organizations usually employ life cycle assessment (LCA), as an environmental assessment approach, to address resource utilization issues. However, it primarily, depends on the inventory of potential pollutant flows through various environmental areas (including the geo-environment), and the assessment of related ecological effects (ISO, 2006). A drawback with LCA approaches, nevertheless, is that they solely consider the secondary effects associated with plastic use, especially resource depletion, energy consumption, or chemical contaminant greenhouse gases, instead

Table 1 Primary chemicals consumed by plastics (Groh et al., 2019; UNEP, 2018; WEF, 2016)

Primary chemicals	Total global consumption (MMT)	Consumed by plastics (%)	Consumed by plastics (MMT)	Common additives/ compounds
Ethylene	113.18	84	95.13	Stabilizers, antioxidants, colorants
Benzene	39.67	85	33.52	Fillers, lubricants, surfactants
Propylene	74.90	82	61.66	Heat stabilizers, lubricants, colorants
Xylenes	42.89	88	37.62	coatings, adhesives, and varnishes
Butadiene	9.28	94	8.75	Household goods, plastic toys
Methanol	41.86	10	4.19	Laminates, bottles
Chlorine	56.21	42	23.55	Pipes, packaging

MMT—Million metric tonnes

of just plastic as a pollutant. Neither a database of marine plastic flows nor a description of variables for assessing the detrimental impact of plastics on ecosystems or human health are provided by LCA (Magnusson et al., 2016).

Therefore, materiality is a crucial technique to combine with circularity and life cycle assessment (LCA) to address marine plastic leaks (Medellin Declaration, 2016). Although there is currently no standard method for conducting such an evaluation, mapping various plastic applications might be accomplished with ease using a straightforward four-level qualitative grading system.

3.3 *Marine Plastic Footprint*

The IUCN had been working with Environmental Agency to establish the Marine Plastic Footprint technique. It extends on the listing of different types of plastic losses which take place during an organization or product's life cycle (Pirc et al., 2016). The objective of the Oceanic Plastic Footprint concept is to further expand the understanding concerning marine plastic leakage and encourage advancements in the integration of plastic leakage into the LCA framework. It integrates the ideas of plastic materiality (the ratio of functionality/added value to potential implications), plastic circularity (the proportion of material recovered or replenished in biological cycles), and plastic leakage (inventory) as can be observed from Fig. 3. Gross assumptions underlie plastic fate and loss rates. Both inadequately handled plastic trash and microplastics from various sources are included in the suggested methodology. For major industries, it offers established emission factors. These calculations include microplastics from four main categories (textile fibres, rubber dust, cosmetics, and

Table 2 Different methodological descriptions used by various companies (Saidani, 2017; Woods et al., 2016)

Methods	Inventory	Scope	Applications
Life cycle assessment (LCA)	Evaluates environmental impacts of a product from cradle to grave	Carbon footprint, water usage, energy consumption, waste generated	Assessing the impact of a plastic bottle from production to disposal
Material flow analysis (MFA)	Analyzes the flow of materials through a system or region	Total plastic production, consumption rates, waste generation	Tracking plastic usage in a city or country
Carbon footprint analysis	Measures greenhouse gas emissions associated with plastic use	CO_2 emissions per unit of plastic produced/consumed	Estimating emissions from plastic packaging
Waste composition analysis	Studies the types and amounts of plastics in waste streams	Types of plastics (PET, HDPE, etc.), weight, volume	Analysing municipal solid waste for plastic content
Ecological footprint	Assesses the demand for natural resources against the planet's ability to regenerate them	Biocapacity, ecological deficit/ surplus	Evaluating the sustainability of plastic consumption in a region
Social life cycle assessment (S-LCA)	Evaluates social impacts of plastic production and use	Labor conditions, community health, stakeholder involvement	Assessing the social implications of plastic manufacturing in developing countries
Input–output analysis	Examines economic data to understand environmental impacts across sectors	Economic transactions, environmental impacts per industry	Evaluating the economic footprint of the plastic industry
Plastic waste generation model	Predicts future plastic waste based on consumption patterns	Projected waste volumes, recycling rates	Modelling plastic waste for a specific region over time

fabrication pellets) in addition to macro plastics which leak from poorly handled waste (Everaert et al., 2018).

The purpose driving the Oceanic Plastic Footprint is to expand understanding of marine plastic leakage while promoting innovations in integrating the concept of plastic leakage within the LCA framework. It adopts an extensive life-cycle approach, addressing about waste collection throughout and following consumption of the item, as well as extra costs such as the release of microplastic while product use, regular consumption, or transportation (Verschoor et al., 2016). Thus, this methodology not only considers the end life product of plastics, but also includes the leakage from various types of microplastics (Fig. 3).

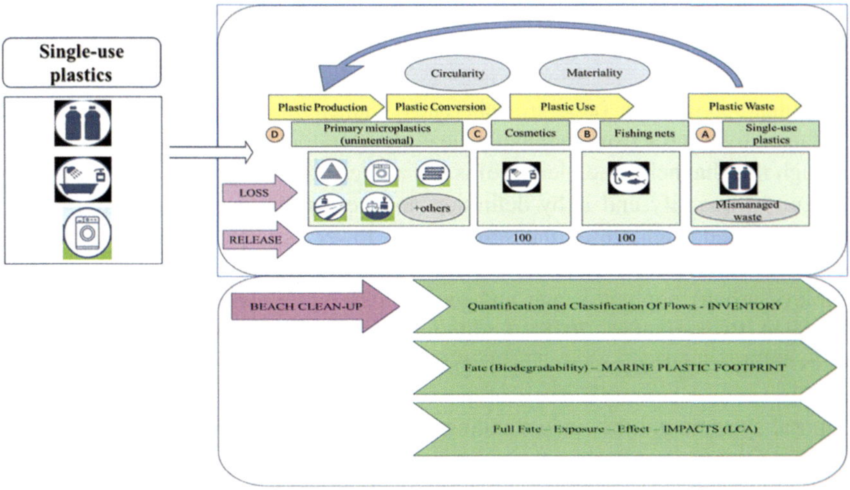

Fig. 3 Generic system map and calculation methods for the marine plastic footprint for different categories of leakage (Boucher et al., 2019)

In general, modelling approaches include quantifying the waste plastics discarded in the marine environment that is obtained from several secondary databases. These discharge rates will be greatly affected by a range of variables, such as social norms (similar to littered procedures), environmental variables (for instance, the influence caused by wind or rain on the migration of waste from dumpsites), spatial particular characteristics (e.g. its proximity from beaches and streams), and financial factors (similar to the material's residual monetary value). The quantity of plastics that enter the ocean, is determined by the volume of macroplastic waste generated at the specified geography location and multiplied by the defined release rate (RR), resulting in an overall leakage as shown in Eq. (1) (CPA, 2014; Rossi & Heine, 2014).

$$\text{Plastic Leakage}_{\text{waste}} = \text{Marine Plastic Footprint} \times \text{Release Rate} \tag{1}$$

$$\text{Release Rate} = \text{RR}_{\text{factor}} \times \text{RR}_{\text{high}} \tag{2}$$

where, RR factor denotes function of the residual value of plastic which is being discarded into the oceans. Accordingly, the probable Marine Plastic Footprint of a certain location can be graphically represented on a digital map if the above equations are computed utilizing the most accurate information available and uploaded to a GIS network (Wagner et al., 2018). The rate of photo, thermal and mechanical degradation for different grades of plastics is mostly influenced by geographical location, with oxygen levels having an essential impact on the process (the rate of degradation at the ocean bottom will be slower than in terrestrial, UV-exposed places). Photodegradation and chemical degradation are the main processes accelerating the disintegration of polyethylene. The Marine Plastic Footprint may thus be computed using the following formula, where 100 years is used as the standard (Lebreton et al., 2017).

Evaluating littering rates are inherently challenging. Although some may be explained by figures from municipal cleaning operators, the portion that "comes in through the channels" (i.e., leakage) is not addressed. This proportion is extremely difficult to quantify and is by definition not measurable. Recent research studies in Switzerland, highlighted the substantially lower estimate for the littering rate to <0.1%. In fact, littering patterns differ depending on the geographical locations (rural vs. urban), population age, and gender in the nations where research have been compiled (Everaert et al., 2018). Additionally, it may also include the proximity and accessibility of garbage cans. This suggests that there may be an abundance of variation based on a country's levels of growth and development, the state of its waste disposal system, and the understanding and consciousness of its people regarding littering habits.

In order to have more primary data available for deducing any patterns, as well as it recommends that the first step in quantifying littering rates may differ by nation be the collection of quantitative data on the mass of plastic litter collected through surveys and investigations. Keeping in mind that the primary objective of this methodology was plastic leakage in the marine environment, though leakage in additional ecological categories (including groundwater or air) is also a significant concern, this is above the boundaries of this work. This gap is being filled by other ongoing programs, such as the Plastic Leak Project. In low and mid-level countries, greenhouse gasses and potentially harmful pollutants are released when garbage is openly burnt without adequate fume treatment. Burned garbage can cause the development of microplastics but does not add to the macroplastic debris that can leak into the environment. The generic loss rates obtained from various plastics sources has been summarized in Table 3.

3.4 Plastic Scorecard

The Plastics Scorecard will promote the creation and application of plastics that employ safer chemicals at every stage of polymer manufacture and in the additive selection process. It evaluates the primary, intermediate, and monomers—the three main chemical inputs used in the production of polymers—in addition to estimate the vulnerabilities associated with it. BizNGO created the Plastics Scorecard to minimize the use of highly hazardous chemicals throughout plastic the production and products (CPA, 2014). The harmful chemical footprint of plastics can be assessed on a scale of 0 (most hazardous) to 100 (most benign) depending on the concentration and percentage of high-concern substances present in items made from plastic. Scores ranging from A+ to F symbolize the footprint as can be seen from Fig. 4. The quantity and weight (or proportion weight) present in a typical plastic product, such as a plastic "rubber" or a protective covering for the screen of a computer, can be measured by

Table 3 Generic loss rates for different plastic sources (Boucher et al., 2019; Linder et al., 2017)

Source	Sub-FU	Unit	Equivalent per capita/year	Equivalent per capita	Quality of data
Textiles	1 kg of textiles washed	Mass of textiles washed over life cycle	Hypothesis: 3.5 kg per wash 165 wash cycles, household of 2.2 persons, share of synthetic textiles = 45%	15	Average
Tyres (car)	1 km driving	Distance driven over life cycle	From 0.23 to 4.7 kg/year, with a global average of 0.81 kg/year—includes synthetic and natural rubber	430	High
Tyres (truck)	1 km driving	Distance driven over life cycle	From 0.23 to 4.7 kg/year, with a global average of 0.81 kg/year—includes synthetic and natural rubber	430	High
Marine paint	1 tkm marine transport	Distance driven over life cycle	loading rate (for product)	Global assumption based on world paint production	4	Low
Road marking	1 kg paint	Amount of paint used over the life cycle of a product/company or in a country	Calculation based on world use	0.1	Low

FU: Functional Unit, Quality of Data—High—several sources of data, narrow range, mechanism is well understood, Average—one of only a few sources of data or wide range of values reported, Low—mechanism is not well understood and lacks data

determining the chemical footprint of a plastic product. It has been reported that the chemical footprint of polyolefin IV bags is significantly lower than that of PVC/DEHP (di(2-ethylhexyl) phthalate) IV bags.

The Plastics Scorecard is a tool for enhancing plastics' both human and environmental efficiency. It is not intended to evaluate effects over a plastic product's whole life cycle or to compare other material kinds, such as aluminium and plastics. However, it gives information about the chemical footprint of polyethylene in addition to how they differ from one another in this aspect (Wagner et al., 2018). Additionally, the raw material feedstocks—such as corn, sugar cane, natural gas, or crude oil—for the polymer inputs are not taken into account by the Plastics Scorecard software version 1.0. This version will evaluate only the finished product's and manufacturing's chemical footprint. Different types of feedstocks for raw materials will not be considered used for the evaluation (UNEP, 2018). A buyer or designer may first screen for feedstock sources that raise issues, such as the use of genetically

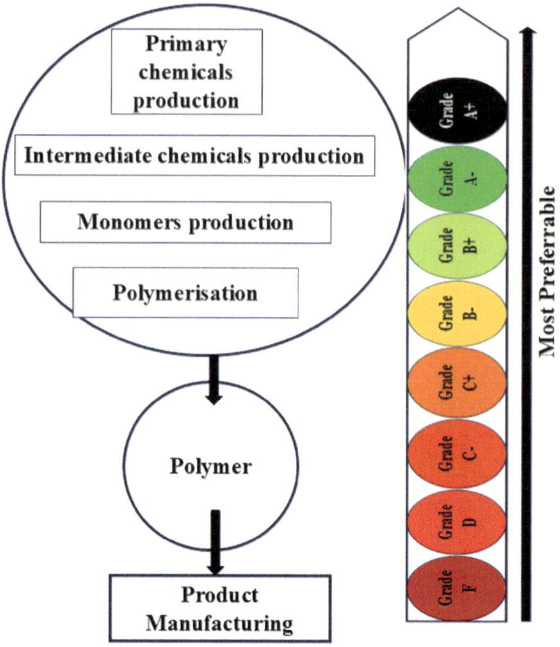

Fig. 4 Generic system map and calculation methods for the marine plastic footprint for different categories of leakage (Rossi & Heine, 2014)

modified organisms (GMOs) in the field or the usage of food crops to make plastics, and then optimize based on chemical footprint (Jambeck et al., 2015).

4 National Guidance for Marine Plastic Hot Spotting and Shaping Action

A strategic framework called National Guidance for maritime Plastic Hot spotting and Shaping Action was established to identify and deal with the main causes of plastic pollution in oceanic habitats (UNEP, 2018). It offers a systematic approach for businesses, communities, and governments to identify important "hotspots" where waste plastics accumulates up, particularly in marine and coastal areas. Through the use of data-driven techniques including waste audits, satellite images, and marine debris surveys, the guidelines assist in identifying high-impact regions where plastic waste management initiatives can provide the greatest benefits. These hotspots could include areas close to industrial or tourist sites, river mouths, urban beaches, or areas where plastic leakage is most common or waste management systems are insufficient.

A multi-stakeholder approach is recommended by the regulations in order to develop focused actions and solutions for minimizing the production of plastic waste. In order to transform into reality successful initiatives like enhanced separation of waste, recycling initiatives, and sustainable manufacturing procedures, municipal governments, the commercial sector, non-governmental organizations, and communities need to collaborate together. In an attempt to protect ecosystems from plastic pollution, the national plan additionally emphasizes the significance of strengthening regulatory frameworks, such as extended producer responsibility (EPR) applications, restrictions on single-use plastics, and the construction of marine protected areas. Initiatives for behavioural change, education, and general awareness activities are also crucial in forming coordinated action against marine plastic waste (PWM, 2018).

These guidelines urge innovative approaches such as fostering renewable options, eliminating the use of plastic packaging, and supporting the circular economy by means of reusing plastic, in addition to tackling the underlying causes of plastic pollution (Medellin Declaration, 2016). Innovation technologies such efficient waste management systems, can help in assist minimizing down on plastic leakage. Furthermore, collaboration between nations will ensure that initiatives align with global frameworks including sustainable development goals (SDGs). Therefore, limiting the leakage of plastics into the oceans, encouraging sustainable resource use, and preserving the health of marine ecosystems for future generations are the ultimate objectives of this mission.

5 Conclusions

This chapter emphasizes about the one of the biggest challenges associated with the reducing plastics' footprint and related methodologies. Exposure to an extensive variety of chemicals of high concern during manufacturing, distribution, transportation and disposal causes serious threat to the industrial workers as well as to the environment globally. Plastics Scorecard can be utilized as an effective tool for evaluating the emissions occurring at each stage of plastic production. It also can be concluded that if circularity, waste generation models, and social life cycle assessment are used synergistically in identifying various scenarios, it will be more effective in aiding the environmental impacts. Moreover, stakeholders, industrialists, environmental planners, and decision makers should encourage for utilizing the leakage-based metrics instead of focusing on only recycling rates. Although, it has been reported that it is more difficult to model plastic leakage from inadequately managed waste materials as it incorporates behavioural (such as littering behaviours), climatic and geographic (such as proximity to beach and rivers) factors. However, the plastic footprints can be lowered with the reducing the utilization of toxic chemicals and increase the use of greener chemicals which will not help the health conditions for the workers but will also provide greener jobs, and generate new markets for different organizations.

References

Ahmed, M., Bashar, I., Alam, S. T., Wasi, A. I., Jerin, I., Khatun, S., & Rahman, M. (2021). An overview of Asian cement industry: Environmental impacts, research methodologies and mitigation measures. *Sustainable Production and Consumption, 28*, 1018–1039. https://doi.org/10.1016/J.SPC.2021.07.024

Akhtar, R., Sirwal, M. Y., Hussain, K., Dar, M. A., Shahnawaz, M., & Daochen, Z. (2022). Impacts of plastic waste on the coral reefs: An overview. In *Impacts of plastic waste on the marine biota* (pp. 239–256). Springer. https://doi.org/10.1007/978-981-16-5403-9_13

Ameen, H. M., Rohini, P., Madhu, M., Rajani, V., & Jayadev, A. (2024). Impact of microplastic pollution in seagrass ecosystem. *Current World Environment, 19*(1), 46–56. ISSN: 0973-4929. https://doi.org/10.12944/cwe.19.1.5

Borrelle, S. B., Ringma, J., Law, K. L., Monnahan, C. C., Lebreton, L., McGivern, A., Murphy, E., Jambeck, J., Leonard, G. H., Hilleary, M. A., Eriksen, M., Possingham, H. P., De, Frond H., Gerber, L. R., Polidoro, B., Tahir, A., Bernard, M., Mallos, N., Barnes, M., Rochman, C. M. (2020). Predicted growth in plastic waste exceeds efforts to mitigate plastic pollution. *Science, 369*(6510), 1515–1518.

Boucher, J., Dubois, C., Kounina, A., & Puydarrieux, P. (2019). *Review of plastic footprint methodologies: Laying the foundation for the development of a standardized plastic footprint measurement tool.* IUCN. https://doi.org/10.2305/IUCN.CH.2019.10.en

Cássio, F., Batista, D., & Pradhan, A. (2022). Plastic interactions with pollutants and consequences to aquatic ecosystems: What we know and what we do not know. *Biomolecules.* https://doi.org/10.3390/biom12060798

Chamas, A., Hyunjin, M., Jiajia, Yang, Q., Tarnuma, T., Jang, J. H., Mahdi, O., Susannah, L., Scott, & Suh, S. (2020). Degradation rates of plastics in the environment. *ACS Sustainable Chemistry & Engineering, 8*(9), 3494–3511.

Clean Production Action. (2014). *Chemical footprint.* http://www.cleanproduction.org/chemical-footprint

Dalui, P., & Jyothi, B. (2021). A study on microplastics: The invincible environmental pollutants. *Pollution Research, 40*(4), 1601–1607. ISSN: 0257-8050.

Emami, N., Baynes, T. M., Kaushik, T., Singh, M., Bhattacharjya, S., Locock, K., & Schandl, H. (2024). Plastic in the Indian economy. A comprehensive material flow analysis. *Journal of Material Cycles and Waste Management, 26*, 3584–3595. https://doi.org/10.1007/s10163-024-02060-z

Everaert, G., Van Cauwenberghe, L., De Rijcke, M., Koelmans, A. A., Mees, J., Vandegehuchte, M., & Janssen, C. R. (2018). Risk assessment of microplastics in the ocean: modelling approach and first conclusions. *Environmental Pollution, 242*, 1930–1938. https://doi.org/10.1016/j.envpol.2018.07.069

Gewert, B., Plassmann, M. M., & Macleod, M. (2015). Pathways for degradation of plastic polymers floating in the marine environment. *Environmental Science. Processes & Impacts, 17*, 1513–1521. https://doi.org/10.1039/c5em00207a

Govani, J., Patel, H., Pujara, Y., Chabhadiya, K., & Pathak, P. (2020). Technological upgradation and advancement for treatment of plastic waste in developing countries. In C. Hussain (Ed.), *Handbook of environmental materials management.* Springer. ISBN: 978-3-319-58538-3.

Groh, J., Backhaus, T., Carney-Almroth, B., Geueke, B., Inostroza, P. A., Lennquist, A., Leslie, H. A., Maffini, M., Slunge, D., Trasande, L., & Warhurst, A. M. (2019). Overview of known plastic packaging-associated chemicals and their hazards. *Science of the Total Environment, 651*, 3253–3268. https://doi.org/10.1016/j.scitotenv.2018.10.015

International Organization for Standardization (ISO). (2006). Management environmental—Analyse du cycle de vie—Principes et cadre. ISO 14040:2006(fr).

Jambeck, J. R., Geyer, R., Wilcox, C., Siegler, T. R., Perryman, M., Andrady, A., Narayan, R., & Law, K. L. (2015). Plastic waste inputs from land into the ocean. *Science, 347*, 768–771.

Kaladharan, P., Singh, V. V., Asha, P. S., Edward, L., & Sukhadane, K. S. (2020). Marine plastic litter in certain trawl grounds along the peninsular coasts of India. *Marine Pollution Bulletin, 157*. https://doi.org/10.1016/j.marpolbul.2020.111299

Kühn, S., Rebolledo, E. L. B., & Franeker, J. A. V. (2015). *Deleterious effects of litter on marine life: Marine anthropogenic litter*. Springer Open. https://doi.org/10.1007/978-3-319-16510-3_4

Kumar, R., Sinha, R., Rakib, M. R. J., Padha, S., Ivy, N., Bhattacharya, S., Dhar, A., & Sharma, P. (2022). Microplastic pollution load in Sundarban delta Bay of Bengal. *Journal of Hazardous Materials Advances, 7*. https://doi.org/10.1016/j.hazadv.2022.100099

Lebreton, L. C., Van Der Zwet, J., Damsteeg, J. W., Slat, B., Andrady, A., & Reisser, J. (2017). River plastic emissions to the world's oceans. *Nature Communications*, 15611. https://doi.org/10.1038/ncomms1561

Linder, M., Sarasini, S., & Loon, P. (2017). A metric for quantifying product-level circularity. *Journal of Industrial Ecology, 21*, 545–558. https://doi.org/10.1111/jiec.12552

Magnusson, K., Eliasson, K., Fråne, A., Haikonen, Olshammar, M., Stadmark, J., & Hultén, J. (2016). *Swedish sources and pathways for microplastics to the marine environment*. IVL Svenska Miljöinstitutet.

McKinsey Center for Business and Environment and Ocean Conservancy. (2015). Stemming the tide. In *Land-based strategies for a plastic—Free ocean*. https://www.mckinsey.com/business-functions/sustainability/our-insights/stemming-the-tide-land-based-strategies-for-a-plastic-free-ocean

Medellin Declaration. (2016). *Medellin declaration on marine litter in life cycle assessment and management*.

Napper, I. E., & Thompson, R. C. (2016). Release of synthetic microplastic plastic fibres from domestic washing machines: Effects of fabric type and washing conditions. *Marine Pollution Bulletin, 112*, 39–45. https://doi.org/10.1016/j.marpolbul.2016.09.025

NITI Aayog. (2022, May). *Report on alternative products and technologies to plastics and their applications*. https://www.niti.gov.in

Pirc, U., Vidmar, M., Mozer, A., & Kržan, A. (2016). Emissions of microplastic fibers from microfiber fleece during domestic washing. *Environmental Science and Pollution Research, 23*, 22206–22211. https://doi.org/10.1007/s11356-016-7703-0

Plastic Waste Management Rules. (2018). Ministry of Environment, Forest and Climate Change. Retrieved October 25, 2024, from https://cpcb.nic.in; ASTM D3172-13. (2013). *Standard practice for proximate analysis of coal and coke*. American Society for Testing and Materials (ASTM). https://www.astm.org/d3172-13r21e01.html

Pujara, Y., Pathak, P., Sharma, A., & Govani, J. (2019). Review on Indian municipal solid waste management practices for reduction of environmental impacts to achieve sustainable development goals. *Journal of Environmental Management, 248*, 109238.

Rossi, M., & Heine, L. (2014). *The green screen for safer chemicals: Evaluating flame retardants for TV enclosures*. Clean Production Action. http://wwwbizngo.org/static/ee_images/uploads/resources/EvaluatingFlameRetardants_GreenScreenSaferChemicals_2007.pdf

Saidani, M., Yannou, B., Leroy, Y., & Cluzel, F. (2017). How to assess product performance in the circular economy? Proposed requirements for the design of a circularity measurement framework. *Recycling, 2*. https://doi.org/10.3390/recycling2010006

Tejaswini, M. S. S. R., Pathak, P., & Gupta, D. K. (2022a). Sustainable approach for valorization of solid wastes as a secondary resource through urban mining. *Journal of Environmental Management, 319*, 115727. https://doi.org/10.1016/j.jenvman.2022.115727

Tejaswini, M. S. S. R., Pathak, P., Seeram, R., & Ganesh, S. P. (2022b). A comprehensive review on integrative approach for sustainable management of plastic waste and its associated externalities. *Science of the Total Environment*, 153973. https://doi.org/10.1016/j.scitotenv.2022.153973

Tejaswini, M. S. S. R., Pathak, P., Singh, L., Raj, D., & Gupta, D. K. (2023). A novel circular approach to analyze the challenges associated with micro-nano plastics and their sustainable remediation techniques. *Journal of Environment Science Health, Part A, 58*, 7. https://doi.org/10.1080/10934529.2023.2208507

Thushari, G. G. N., & Senevirathna, J. D. M. (2020). Plastic pollution in the marine environment. *Heylion, 6*(8). https://doi.org/10.1016/j.heliyon.2020.e04709

UNEP. (2018). *Marine plastic debris and microplastics—Global lessons and research to inspire action and guide policy change*. UNEP.

Unice, K. M., Weeber, M. P., Abramson, M. M., Reid, R. C. D., Van Gils, J. A. G., Markus, A. A., Vethaak, A. D., & Panko, J. M. (2018). Characterizing export of land-based microplastics to the estuary—Part II: Sensitivity analysis of an integrated geospatial microplastic transport modeling assessment of tire and road wear particles. *Science of the Total Environment*. https://doi.org/10.1016/j.scitotenv.2018.08.30

Verschoor, A., De Poorter, L., Dröge, R., & De Valk, E. (2016). Emission of microplastics and potential mitigation measures. In *Abrasive cleaning agents, paints and tyre wear*. National Institute for Public Health and the Environment (RIVM).

Vivekanand, A. C., Mohapatra, S., & Tyagi, V. K. (2021). Microplastics in aquatic environment: Challenges and perspectives. *Chemosphere, 282*, 131151. https://doi.org/10.1016/J.CHEMOSPHERE.2021.131151

Wagner, S., Hüffer, T., Klöckner, P., & Wehrhahn, M. (2018). Tire wear particles in the aquatic environment—A review on generation, analysis, occurrence, fate and effects. *Water Research, 139*, 83–100. https://doi.org/10.1016/j.watres.2018.03.051

Woods, J. S., Veltman, K., Huijbregts, M. A. J., Verones, F., & Hertwich, E. G. (2016). Towards a meaningful assessment of marine ecological impacts in life cycle assessment (LCA). *Environment International, 89–90*, 48–61. https://doi.org/10.1016/j.envint.2015.12.033

World Economic Forum (WEF). (2016). The new plastic economy. In *Rethinking the future of plastics*. WEF. http://www3.weforum.org/docs/WEF_The_New_Plastics_Economy.pdf

Nanobiotechnology for Reducing Plastic Footprint and Combatting Micro and Nano-plastic Pollution

Abhishek Gupta, John Majok Manyok Thuch, Vranda Aggarwal, and Palak Sharma

Abstract Plastic pollution, particularly from micro and nano-plastics, has become a significant environmental threat, impacting ecosystems and human health due to their persistence, widespread presence, and contributions to an increasing global plastic footprint. Conventional methods face limitations in addressing this issue, but nanobiotechnology offers innovative and sustainable solutions. This chapter examines how nanotechnology and biotechnology can be combined to create advanced tools for the detection, mitigation, and removal of plastic pollutants in air, water, and soil, ultimately reducing the overall plastic footprint. Nanomaterials, such as nanoparticles, nanocomposites, and functionalized nanostructures, enhance the sensitivity of detection, accelerate degradation, and improve filtration. For example, nanosensors enable real-time monitoring of plastic particles in various environments, while functionalized nanoparticles facilitate targeted adsorption and removal of microplastics. Moreover, bioengineered microorganisms and enzyme-immobilized nanoparticles represent novel remediation techniques for breaking down plastics directly within the environment, helping to offset the long-term accumulation of plastics. These strategies not only mitigate immediate pollution but also contribute to a systematic reduction in the total plastic footprint by promoting sustainable cycles of plastic degradation and reuse. By leveraging the unique capabilities of nanomaterials and the precision of biological systems, nanobiotechnology holds the potential to significantly reduce environmental and health risks while advancing global sustainability goals. The future of this field will depend on interdisciplinary research and technological innovation to overcome existing challenges and scale up these solutions for widespread global impact.

Keywords Micro and nano plastic · Nanosensors · Nanobiotechnology · Plastic pollution · Plastic footprint

A. Gupta (✉) · J. M. M. Thuch
Department of Chemical Engineering, Marwadi University, Rajkot, Gujarat, India
e-mail: abhishek.gupta@marwadieducation.edu.in

V. Aggarwal · P. Sharma
School of Engineering and Applied Sciences, Bennett University, Greater Noida, Uttar Pradesh, India

P. Pathak et al. (eds.), *Plastic Footprint*, https://doi.org/10.1007/978-981-96-6362-0_6

1 Introduction

1.1 Overview of Micro/Nano Plastic Pollution

One of the biggest environmental challenges we face today is pollution from plastics. Microplastics and nanoplastics, in particular, pose significant risks due to their small size, which often renders them nearly invisible in various ecosystems. Microplastics are defined as plastic particles ranging in size from 1 micron to 5 mm, while nanoplastics are particles smaller than 1 micron (Peng et al., 2020). The term "plastic footprint" provides a useful lens to understand the scale of the problem. It refers to the total amount of plastic consumed and discarded by individuals, organizations, or regions, offering a direct link to environmental degradation. Over 300 million tons of plastic are produced annually, with a significant portion comprising single-use plastics like packaging and disposable items. Improper waste management and low recycling rates amplify the issue, as plastics infiltrate terrestrial, aquatic, and atmospheric ecosystems. As they break down, they contribute to the accumulation of micro/nanoplastics, demonstrating the interconnected nature of plastic consumption, waste, and pollution. Reducing the plastic footprint through sustainable practices and biodegradable alternatives is essential to mitigate these cascading effects. Microplastics and nanoplastics can be classified based on their origins (Peng et al., 2020).

Primary sources—It also encompasses the planned or the directly discharged micro/nanoparticles such as microbeads in cosmetic products, synthetic fleece in clothes and particles employed in manufacturing of plastics (Fig. 1).

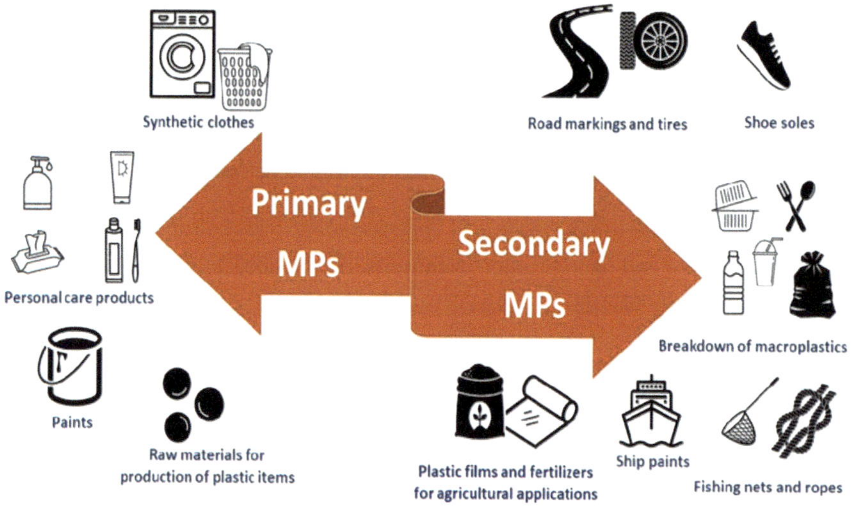

Fig. 1 Primary and secondary sources of micro plastic (Acarer, 2023)

Secondary sources—Secondary micro/nanoplastics are further divided into macroplastics such as bottles, bags, and packaging material that degrade into micro/nanoparticles due to UV radiation, abrasion, and microbial action.

Micro/nano plastics offer a universal perspective on the pervasiveness of plastics in both usage and waste, demonstrating their persistence and non-biodegradability. These particles have infiltrated terrestrial, aquatic, and atmospheric environments worldwide, with high concentrations observed in specific settings:

(a) **Oceans and Water Bodies**—The marine environment is a major hotspot for micro/nanoplastics, as particles settle in coastal, stream, and deep-sea deposit areas. These plastics enter the marine food chain, being ingested by both micro and large marine life such as zooplankton and whales. They also attract pollutants that adhere to the plastic surfaces, creating additional ecological risks.
(b) **Freshwater Systems**—Micro and nanoplastics are found across forests, fields, oceans, and various types of wetlands. They are transported from land into rivers, lakes, and streams, where they accumulate in freshwater environments. This pollution threatens the development of freshwater organisms and disrupts the natural balance of these ecosystems.
(c) **Soil**—Micro and nanoplastics have been identified in soils across diverse land-use areas, including agricultural, urban, and natural landscapes. These plastics can be ingested by earthworms and other soil invertebrates, reducing soil fertility and altering its biological functions (Padervand et al., 2020).
(d) **Air**—Micro and nanoplastics are also dispersed through the air, carried by wind over long distances. Studies confirm the presence of these particles in urban air samples, suggesting they may act as carriers for other pollutants under certain conditions.

1.2 Impact of Micro and Nanoplastic Pollution

Despite their small size, microplastics and nanoplastics (less than five millimeters and even down to nanometers in scale) represent significant threats to ecosystems, environmental health, and human well-being. Preventing pollution from these plastics is essential for multiple reasons.

1.2.1 Environmental Impact

The effects of micro/nanoplastics are widespread in a wide range of biomes, starting from marine and reaching terrestrial ones. The ingestion of the micro and nanoplastics in oceanic ecosystem can have perils on the parts of fishes, sea birds and marine mammals such as causations of physical harm, chronic blockages of their body tracts and toxic implications. Furthermore, the micro/nanoplastics can be embedded in marine sediments, contributing to change in sediment chemistry, destruction of benthos communities, and inhibition of nutrient cycling. This accumulation also has a potential danger to filter feeders like bivalves and corals that may likely ingest micro/nanoplastics together with their nutrient intake, thereby resulting in bio accumulation, biomagnification of associated pollutants. Moreover, micro/nanoplastics have the capability of carrying invasive species and environmental pollutants, thus, serve as carriers of invasive species, as well as potential agents for spreading hazardous chemicals within ecosystems (Musa et al., 2024).

In terrestrial system, accumulation of micro and nanoplastics can result into polluted soil resource, reduced soil fertility, plant and microbial health. The specific threats that farming soils are exposed to include invasion of micro and nanoplastics, which will have detrimental implications on the production of food, the supply of food and the services offered by soil ecosystems. Further, microplastics may leach out nasty substances into the environment and thereby pollute soil and water that ultimately are lethal to aquatic and terrestrial life forms. Therefore, the mitigation of micro/nanoplastics pollution requires critical and immediate attention to conserve the quality of environment, enhance the protection of ecosystems and promote effective land and water management techniques.

1.2.2 Ecological Integrity

Micro and nanoplastic pollution pose a strong risk for ecological health due to the interference with ecological processes, change in physical features of the ecosystem and interaction of species. Feeding studies have shown that micro/nanoplastics can affect the feeding behavior and intake rate of food of resource bacteria and zooplankton and may also reduce the reproductive rates of fish, which may lead to population reduction and shift in the community structure of aquatic organisms (Nzeyimana et al., 2025). Furthermore, it was estimated that micro/nanoplastics could also create a suitable microbial surface for microbial adaption and proliferation of pathogenic bacteria and toxic alga in water, making water quality worse and threatening the survival of organisms in an aquatic ecosystem (Musa et al., 2024).

In terrestrial ecosystems several macros and mesofauna including earthworms and microarthropods which are involved in the decomposition of organic matter; modification of particles, root penetration, aeration, and nutrient cycling are influenced by micro/nanoplastics. Ingesting the micro/nanoplastics may lead to physical injury, gut blockage, nutrient deficiencies, and ultimately alter soil fertility and ecosystem service. Furthermore, micro/nanoplastics can concentrate in sediments, thus exerting impacts on benthic and biological communities within the sediment, and the food chain and ecosystem balance.

Ecosystemic intercalation is important for the conservation of biological diversity, and stability and for the provision of ecosystem services for human welfare. Thus, the prevention and control of micro/nanoplastic pollution are essential for effective protection of ecosystems and species, which will preserve the essential function of ecosystem services.

1.2.3 Public Health Issues

The pollution with Micro/nanoplastics is a threat to public health and has impacts for human health and welfare. Intake of micro/nanoplastics due to the consumption of contaminated food, water, and breathing in contaminated air led to the physical, chemical, and microbial, in imminently threatening health. Harmful micro and nano plastics that have a cutting or abrasive edge can also cut up the gastrointestinal tract, causing inflammation, ulcers, or gastrointestinal diseases. Furthermore, micro/nanoplastics become vectors and receptors for absorbing and accumulating toxic pollutants, including heavy metals, pesticides, and endocrine disturbing chemicals, which induce toxic impact on the exposed person (Pelegrini et al., 2023).

Also, micro/nanoplastics act as vectors transporting microbial pollutants like bacteria and viruses and, therefore, pose increased potential of water- and foodborne diseases. From the current findings results, micro/nano-plastic pollution has longer residual effects that continue to elicit community health concerns especially children, pregnant women, and other groups with sensitive susceptibilities, including groups from impoverished areas that continue to experience localized pollution increases.

Consequently, there is a need to mitigate the problem of micro/nanoplastic to enhance public wellbeing, minimize the impacts on the environment and advance sustainable development. Thus, preventing the source, spreading and influence of micro/nanoplastics release, one can save ecosystems, support biodiversity, and enhance the health and quality of life of the present generation and those to come. It is recommended that micro/nanoplastic pollution issue requires the cooperation of government, industry, academia and civil society to find out an all-round solution and make the world a safer place to live for all.

1.3 Role of Nanobiotechnology in Environmental Enhancement

As the world grows increasingly aware of the urgent need to address micro- and nanoplastic pollution, nanobiotechnology emerges as a promising strategy for environmental conservation. Leveraging the advantages of nanomaterials in biological applications, nanobiotechnology offers innovative methods to detect, monitor, and manage micro- and nanoplastic pollution across diverse environments (Bhattacharjee & Roy, 2023).

At its core, nanobiotechnology combines principles from nanotechnology, biotechnology, and environmental science to address environmental challenges. Key properties, such as size-dependent behavior, surface chemistry, and unique optical and magnetic characteristics, enable material scientists to develop new materials and devices that can interact with biological systems at the nanoscale. Additionally, by incorporating enzymes, antibodies, and beneficial microbes, researchers can create biosensitive nanomaterials and nanodevices that are novel, highly specific, efficient, and biodegradable (De & Goswami, 2022).

Nanobiotechnology has numerous uses when it comes to micro/nanoplastic pollution in that it can be used for the identification and characterization of micro/nanoplastics on the one hand, and for the enhancement of pollution removal techniques on the other. All these research threads are integrated in interdisciplinary research to develop new paths to solve the multifaceted issues related to micro/nanoplastic pollution with the focus on the conservation of ecosystems, protection of human health and sustainable development.

In the following chapters for this series, specific with micro/nanoplastic contamination, principle of nanobiotechnology and recent inventions or possibility of using nanobiotechnology for minimizing micro/nanoplastic contamination will be discussed. In this context, the application of an innovative multidisciplinary platform encompassing nanotechnology, biotechnology, and environmental science shall help to delineate the prospect of a safer future with minimisation of micro/nanoplastic contamination risk.

2 Nano Sensors and Their Role in Detecting Micro/Nano Plastics in Various Environments

A nanosensor is an analytical device which can quantify properties at the molecular level including physical, chemical and biological. In the current pollution by micro and nanoparticles including micro/nanoplastic pollution, nanosensors are useful tools for identifying and measuring plastic particles in water, soil and even in the open air.

One of the most significant benefits of a nanosensor is that it has both high sensitivity and high selectivity because of nanofeatures of the material. For instance, nanospheres, nanofibres and nanotubes possess massive surface areas and volume ratio thus the electrical, optical, or even mechanical characteristic pending the usage they are to be subjected to makes them strategic choices in fabricating sensing platforms. By modifying these nanomaterials with specific receptors or ligands, it becomes possible to selectively detect and accommodate micro/nanoplastics from other interferences in real environments.

Thus, nanosensors can be employed in checking the level and uniformity of microplastics in rivers, lake, and ocean waters. Therefore, real-time detection of microplastic concentrations and spatial distribution patterns can be obtained when deploying low-cost and highly sensitive nanomaterial-based sensors onboard autonomous underwater drones or buoy systems. In the same way, nanosensors successfully detect microplastics in the samples of soil and sediments that well describe the presence and effects of the pollution in terrestrial environments.

Micro/nanoplastics biosensing with biosensing elements like enzymes, antibody, or nucleic acids offers a potential way for a fast and effective discovery of the testing material. These biomaterial elements can be designed to capture specific type of plastic particles based on the structure and hence enhancing the sensitivity as well as the selectivity of the sensor.

An example on how the biosensors for detect micro/nanoplastics is through binding aptamers to target molecules aptamers are small single-stranded DNA or RNA molecules. Hence, via systematic evolution of ligands by exponential enrichment (SELEX), aptamers that accommodate surface characteristic or chemical configurations of micro/nanoplastics can be selected. These aptamers can be immobilized on sensor surfaces that can capture and detect plastic particles in the environment samples.

There are also enzyme biosensors for micro/nano plastic detections which work by employing enzymatic reactions for signal magnification and identification. Such enzymes as peroxidase or alkaline phosphatase can become involved in the conversion of chromogenic or fluorescent substrates in the presence of micro/nanoplastics, which can be quantified with optical or electrochemical methods. This enzymatic amplification enhances the sensitivity of the sensor and allows a fast and selective detection of the plastic particles in models of an environmental matrix.

3 Challenges and Advances in Nanotechnology-Based Detection Methods

Nanotechnology based approaches towards micro/nanoplastic detection have a lot of promise however to achieve this several hurdles need to be overcome. Another is that samples from environment are gross and may be laden with interferences that can inflect on the performance of the sensor as well as the quality of results obtained. Hence, establishing sound detection and quantification methodologies that would enable the detection of micro/nanoplastics beyond noise are still important (Claessens et al., 2013).

Another consideration is the standardization or validation of detection methods, so that they will be more accurate and can easily be repeated and compared with those methods employed by another laboratory or centre. Prescribing reference standards and procedures in sensor calibration and result validation is possible to ensure that data come from similar methods, procedures, and qualities and therefore can be compared with other laboratories.

Progress in the fields of nanomaterial fabrication, engineering, and improvement of the sensor system were the main factors that led to the development of micro/nanoplastic detection using nanotechnology. Increasing complexities and innovations in new Nanomaterials with improved sensitivity, selectivity and stability are enhancing the potential of nanosensors and biosensors for accurate and precise detection of plastic particles. Furthermore, miniaturization and integration technologies are pushing the fuss of portable and transportable sensor system for in-situ measurement of micro/nanoplastic concentration in water and other matrices (Claessens et al., 2013).

They highlighted that nanobiotechnology presents potential strategies to detect and monitor micro/nanoplastics in different environments by designing innovative nanosensors and biosensors. Managing factors like sample complexity and standardization is important to fully harness the capacity of nanotechnology-based detection mechanism and get enhanced understanding to the realities of micro/nanoplastic pollution. Nanobiotechnology assures the sustainability of micro/nanoplastics in the environment by taking advantage of nanomaterials and biorecognition elements.

3.1 Application of Nanomaterials to Fix Micro/Nano Plastics in Soil, Water and Air

Microplastics and nanoplastics are now identified as emergent pollutants affecting diverse environments and representing potential hazards towards biota and people. This appears to need unique approaches to practically dismantle and clear them. It is important to present nanomaterials as one of the most powerful tools to combat micro/nanoplastic pollution because of their great characteristics. In this section, we will explore the possibilities of nanomaterials in the removal of micro/nanoplastics from soil, water, and air, opportunities for functional nanoparticles in the adsorption of microplastics, and the possibilities of nanocomposites for separate plastic particle filtration with high efficiency (Table 1).

Table 1 Types of nanoparticles used for the removal of micro/nanoplastics in respective environmental compartments (Allouzi et al., 2021; Youssef et al., 2013; Kathiriya et al., 2021)

Sl. No.	Nanoparticles	Application	Environmental components
1	Titanium dioxide nanoparticles	Photocatalytic degradation	Water
2	Iron oxide nanoparticles	Magnetic assisted removal	Water, soil
3	Silica nanoparticles	Adsorption and filtration	Water
4	Zinc oxide nanoparticles	Photocatalytic degradation	Water
5	Carbon-based nanoparticles (such as graphene, carbon nanotubes)	Adsorption and filtration	Water, soil, air
6	Polymer nanoparticles	Enzymatic degradation enhancement, adsorption	Water, soil
7	Metal nanoparticles (e.g. silver, gold)	Antibacterial properties, catalytic degradation	Water, soil
8	Quantum point	Detection of fluorescent labels	Water, soil
9	Cerium dioxide nanoparticles	Generation of reactive oxygen species promotes degradation	Water
10	Magnetic nanoparticles (such as cobalt ferrite)	Magnetic separation and removal	Water, soil

The major properties of nanomaterials include a large SA/VS ratio, variable surface chemistry and a high reactivity that make it possible to apply these materials in the removal of micro/nanoplastics from soil, water and air. There are several applications of nanomaterials in soil remediation which include micro/nanoplastics separation, improvement of the structure of the soil to enable better removal of the particles from contaminated regions (Musa et al., 2024).

In this context, nanoparticles including zero valent iron (nZVI) and titanium dioxide (TiO_2) have been examined because they can adsorb micro/nano-plastics and enhance their degradation via photocatalytic reactions. These nanoparticles can either be imbedded directly into the matrix of the soil or applied as a surface layer that enables the nanoparticles to interact with micro/nanoplastics and consequently be extracted from the soil system. Additionally, based on the ligands or polymers functionalized on nanomaterials, micro/nanoplastics can be selectively incorporated and remediated or extracted from contaminated soils.

In water treatment, nanomaterials serve as the adsorbent, filters, or photocatalysts in the removal of micro- and nanoplastics from the aqueous environment. Micro-nanoplastics have a high adsorption capacity for these nanoparticles because of a large specific surface area and the availability of multifunctional groups on the nanoparticle surface. These nanoparticles can be applied in filters for water treatment or as adsorbent for micro-and-nanoplastic in water bodies such as rivers, lakes or oceans (Peng et al., 2020).

Furthermore, the photocatalytic degradation of micro- and nanoplastics using nanomaterial-based photocatalysts like semiconductor nanoparticles (TiO_2, Zinc oxide) can also be employed under ultraviolet or visible light conditions (Khan et al., 2024). Phosphorylation of micro- and nanoplastics through the photocatalytic degradation mechanism involves activation of nanoparticles that release reactive species on to the surface of micro-/nanoplastic materials causing fragmentation of the materials into smaller particles. This strategy is ecological and cost-effective, specifically for the removal of micro- and nanoplastics from aquatic ecosystems; it cannot be used in wastewater treatment and ocean pollution mitigation (Peng et al., 2020).

Regarding air treatment, nanomaterials can be applied for the filtration of micro-/nanoplastic in the air of urban and industrial areas. Zeolites and metal organic frameworks (MOFs) including their nanoparticles exhibit high adsorption capacity of micro/nanoplastics; these can be incorporated in air filters or employed as aerosol collectors for trapping plastics in indoor and outdoor environment (Mohan et al., 2024). Furthermore, photocatalysts based on nanomaterials can increase the rate of micro/nanoplastics degradation in the air, using sunlight or artificial light to decrease their concentration in atmosphere consequently.

3.2 Functional Nanoparticles for Adsorption and Removal of Microplastics

Functional nanoparticles are therefore considered the most suitable in the adsorption and eventual removal of microplastics from different environmental matrices. Such nanoparticles are subjected to certain surface treatments in order to improve their binding capability, selectivity and efficiency of microplastic particles removal.

(1) **Magnetic nanoparticles**—Polymer or surfactant-coated magnetic nanoparticles employed in the literature for microplastic removal include iron oxide nanoparticles (Fe_3O_4). These have the benefits of easy to deposit and separate with the help of an external magnetic field. The adsorption capacity and selectivity of these materials for microplastics can be improved if functionalized with polymers or surfactants. In the study, magnetic nanoparticles when dispersed in water or soil environments and containing microplastics can be easily adsorbed on to the plastics. This application of a magnetic field can also recycle magnetic nanoparticles and the adsorbed microplastics, making it easier to expunge them from the environment (Mohan et al., 2024).

(2) **Molecularly Printed Polymer (MIP)**—Molecular imprints are synthetic polymers that have high reactivity, shape selectivity, as well as high specificity for target molecules. In the context of microplastic removal, MIPs can be synthesized by the binding cavities of which can be designed depending on the size, geometry, and surface of the microplastics (Gagliardi & Mazzolai, 2015). Herein, these MIP-functionalized nanoparticles can selectively adsorb microplastics in environmental samples and offer a highly effective integration strategy. Due to the design flexibility of the MIPs, the binding site can easily design to target a given type of microplastics enhancing its suitability for use in different environments (Haupt et al., 2011).

(3) **Hydrophilic/Hydrophobic Nanoparticles**—The selectivity is influenced by the choice of the nanoparticles having hydrophilic or hydrophobic layers on the surface. Cationic particles, for instance, silica nanoparticles or cellulose based nanoparticles, selectively adsorb HPMs because of the hydrophobic nature. Hydrophobic NPs or other carbon-based NPs or polymeric NPs, on the other hand, attracted only hydrophilic microplastics. The ability to selectively target identified types of microplastics in environmental samples is attributed to the ability to tune the surface chemistry of nanoparticles. Selective adsorption ability enhances the efficiency of the microplastic elimination process and prevents the General adsorption of other compounds (Reddy et al., 2016).

Functional nanoparticles have benefits in eliminating microplastics due to factors such as high adsorption efficiency, selectivity and are easily manageable. These nanoparticles can also be designed for forms of microplastics and used in many forms of environmental cleanup solutions. Nevertheless, drawbacks like the stability of nanoparticles, its compatibility with human body, and the effect on the environment must be highly considered for functional nanoparticles can remove microplastics efficiently and safely.

3.3 Nanocomposites and Their Effectiveness on the Filtration of Plastic Particles

Nanocomposites are a class of materials which are made up of nanoscale particles combined with a matrix material, which have shown potential for the effective filtering of plastic particles from all facets of the environment. These new materials make use of the inherent characteristics of nanomaterials, to enhance the filtration mechanism, adsorption ability, and selectivity to the plastic particles. In this section, the design concepts, preparation techniques, and potential uses and efficacy of nanocomposites for filtering out plastic debris specifically from the environment are described.

3.3.1 Design Principles of Nanocomposites for Plastic Particle Filtration

To design nanocomposites for the filtration of plastic particles, one must pay considerable attention to the choice of nanoparticles, the type of the polymer matrix and the technology used in the production of the nanocomposite. For effective broader nanocomposites with high adsorption capabilities and surface area to volume ratios the ideal materials for capturing and retaining plastic particles should be nanoscale materials. They include nanoparticles such as metal oxides and carbon-based particles, nanofibres such as carbon nanotubes and nanocellulose and nanosheets such as graphene oxide.

The matrix material delivers the mechanical strength and stiffness to the nanocomposites and must be compatible with the selected nanomaterial for the application. Polymeric matrices including polymer resins, hydrogels and elastomers are employed more often because of their flexibility and ease of manipulation and processing. There is need to make sure that the nanomaterials are compatible with the matrix material in these nanocomposites so that they can uniformly distribute themselves throughout the material and interact well with the other components to enhance mechanical properties and retard particle coalescence.

Preparation methods of nanocomposites depend on the nature of the nanocomposite to be produced and the type of attributes that the final product should possess. The most frequently employed techniques are solution casting, melt blending, electro spinning and in-situ polymerization. These technologies allow for great control over the dispersion of nanomaterials, the specific morphology of the matrix, and the overall structure of the composite to improve filtration results and application performance.

3.3.2 Application of Nanocomposites in Plastic Particle Filtration

Nanocomposites are used extensively to remove plastic particles from numerous environmental matrices such as water bodies, atmosphere, and land. They include surface area to volume ratio, porosity, and adsorption capacity of the materials that enable them to capture and crystalline the plastic particles through physical capture, adsorption or chemical forces (Youssef et al., 2013).

Nanocomposite membranes have a broad application in water filtration to remove plastics, microplastics, and nanoplastic from water resources. These membranes usually have additives made of nanomaterials including graphene oxide, carbon nanotubes or metal oxides incorporated in a polymer matrix. On the one hand, nanomaterials offer large surface area and great adsorption capacity for the plastic particles, on the other, the polymer matrix gives sparse support and strictly regulates the membrane structure. Nanocomposite membranes present high filtering capabilities of more than 99% efficiency for both small and large, dense and low plastic particle concentrations (Padervand et al., 2020).

4 Design Principles of Nanocomposites for Plastic Particle Filtration

Thus, nanocomposites are materials consisting of one or more discrete, discrete nanoscale particles integrated into a base material. Developing nanocomposites for the filtration of plastic particles requires consideration on nanomaterials and matrix materials and then nanomaterial/matrix interactions and structures to improve filtration efficiency (Kamath et al., 2022).

4.1 Selection of Nanomaterials

As the active phase, nano materials present in nanocomposites are beneficial to endorsing the removal of plastic particles. When selecting nanomaterials, properties that must be taken into consideration include the surface areas, porosity, adsorption capacity and chemical activity of the nanomaterial. Natural and synthetic nanomaterials found in nanocomposites used in plastic particle filtration include nanoparticles, such as metal oxides and carbon-based materials, nanofibers including carbon nanotubes, and nanosheets including graphene oxide (Acarer, 2023; David et al., 2020).

(a) **Surface area and porosity**—These characteristics increase the number of active sites to adsorption and trapping of plastic particles and enhances the filtration efficiency.

(b) **Adsorption capacity**—The nanomaterials used in the study must have a very high adsorption capacity to be able to capture and firmly hold the plastic particles in the nanocomposite matrix.
(c) **Chemical reactivity**—Nanomaterials with other surface functional groups or chemicals attached to them, have higher selectivity towards adsorbing specific types of plastic particle thereby enhancing the selectivity of the nanocomposite.

4.2 Matrix Material Selection

The matrix material sustains and substantiates the nanocomposite's construction by encouraging the dispersion and affixation of the nanomaterial. Polymer matrices are versatile, tough; and can be easily combined with nanomaterials (Bhoi et al., 2020). Yet, other matrix materials as ceramics, zeolites, activated carbon, etc., can be applied just for certain purposes.

(a) **Compatibility**: The matrix material should be compatible with the chosen nanomaterial to guarantee that the nanomaterial is well-dispersed over the matrix material and in contact with the latter there is a strong bond within the nanocomposite. This conformity minimises particle agglomeration and increases mechanical characteristics and filter efficiency.
(b) **Mechanical properties**: The matrix material needs to confer enough mechanical reinforcement on the nanocomposite to withstand the stresses and strains experienced through filtration without changing shape or breaking down.
(c) **Chemical stability**: The matrix material should itself be chemically stable under the conditions experienced during the filtration process for the nanocomposite to demonstrate long life.

4.3 Optimization of Composite Material Structure

Since the nanocomposite materials consist of one or more phases which may be characterized by an unconventional crystalline structure and morphology, the filtration efficiency of such materials is influenced by these characteristics (Acarer, 2023; Bhoi et al., 2020). Dispersion, orientation and arrangement of nanomaterials within the matrix materials can be accomplished through several preparation methods including solution casting, melt mixing, electrospinning and in-situ polymerization.

(a) **Uniform dispersion**: This means that nanomaterials should be well distributed in the matrix material to ensure maximum surface area for accommodation of the plastic particles. It also provides uniform dispersion for enhancing the incorporation of nanomaterials and upgraded the filtration efficiency.
(b) **Interfacial interactions**: For avoiding the increase of particle size unstable aggregation and bad mechanical performance must be avoided by enhancing the Interface adhesion action between Nanomaterials and matrix materials. The

nanoproduct properties can be optimised by the method of surface modification and functionalisation to contribute to the creation of strong bonding between the nanomaterial and matrix materials.

(c) **Controllable morphology**: It is therefore possible to control the morphology of the nanocomposites—particularly, particle size, shape and distribution–to enhance filtration efficiency and selectivity.

5 Biodegradation and Bioremediation Strategies

In the present day, there has been increasing cases of scaring sizes of plastics afloat in the environments and Micro-plastics, Nano-plastics are a threat to lives. This emerged as a major issue where conventional treatment processes cannot solve hence the need to explore new biodegradation and bioremediation techniques. In this section we analyse how nanoparticles help microbial degradation of micro/nanoplastics, application of bioengineered nanomaterial for improving the bioprocessing and biodegradation, and prospects and limitations of nanobiotechnological approaches and their ethical imperatives.

5.1 Nanoparticles Help Microorganisms to Degrade Micro/ Nano Plastics

Biodegradation of plastics by microorganisms is a likely solution to the current problem of plastic pollution since the polymers that make up most plastic can be split by enzymes into simpler molecules that are easier to degrade. The nanoparticles can act as enzyme or microbial cell vectors and offer protection to the enzymes or microbial cells against harsh environmental conditions and foster the interaction between microorganisms and plastic particles, which is responsible for the improved microbial degradation of plastics as postulated by David et al. (2021).

(a) **Immobilization of enzymes**—Nanoparticles bear excellent characteristics of fixed plastic degrading enzymes and are thus unique portals for the deployment of these requisite enzymes. These enzymes include lipase, esterase, oxidoreductase are important in breaking plastic polymers into simpler constituent products that are easier to degrade. Such nanoparticles deliver enzymes with comparatively high stability and activity on the surface of the nanoparticles that continue to catalyze the reactions under adverse environmental conditions. The surface chemistry of nanoparticles enables the enzyme adhesion through physical adsorption or attachment of a functional group. It has been identified that the functional groups on the nanoparticles surface can offer the active points for enzyme attachment whereas the surface alterations can be employed to control the interaction between the enzyme and the nanoparticle. The enzyme, when immobilized on the support, maintains its catalytic properties, and substrate selectivity while showing increased stability against denaturation and proteolytic cleavage.

(b) **Increase bioavailability**—Therefore, the bioavailability of micro/nanoplastics to microbial communities is enhanced by nanoparticles. Microbe particles are generally hydrophobic and tend to clump in aqueous environments and discourage microbial desorption. Ultimately, nanoparticles can enhance the fragmentation and distribution of plastic particles and make them favourable for absorption by enzymes and metabolic paths among microbes (Varshney, 2024). Surface modification stabilizes plastic particles through an increased wettability and dispersibility, using hydrophilic coatings or surfactants, efficient inhibition of re-agglomeration and encourages microbial adhesion. Thus, nanoparticles can also transport hydrophobic plastic additives or nutrients which enhance microbial adhesion and metabolism at the interface of water and plastic. However, nanoparticles can affect microbial degradation in the following ways: they can form favourable microenvironments for microbial degradation, protect against inhibitory compounds and stimulate enzyme activity.

(c) **Environmental restoration**—Concerning environmental remediation, nanoparticles provide various solutions in enhancing the growth and activity of plastic degrading microorganisms in the affected sites. Iron oxide, titanium dioxide and carbon-based nanomaterials used by engineers can get and donating electrons to microbes to partake in redox reactions that commend the breakdown of plastics. When functional nanoparticles are employed, they can effectively target and immobilise non-degrading microbial species thus enhancing the rate of the biodegradation process. Indeed, nanoparticles with antibacterial activity can minimize competition and predation by microbes, consequently, microbial population and metabolism in plastics polluted ecosystems. Nanoparticles can also act as vectors for supplying nutrients, electron shuttles or signalling compounds to microbial consortia hence enhancing plastic degradation and the rate of repair information transfer.

(d) **Controlled Release System**—Self-assembled nanoparticles can act as a controlled-release system that will release enzymes, nutrients or microbial cells at the area within the plastic body or the contaminated site. The encapsulation of the enzyme in a nanoparticle carrier allows it to remain intact, and to be slowly released into the solution over a described time frame. Stimuli sensitive nanoparticles may be designed to be sensitive to parameters such as pH, temperature, or redox potential and release, and activate enzymes as a response to the microbial presence. Due to the selectivity applied in the management of enzyme activity and location within the controlled-release systems, the plastic degradation efficiency can be maximized without high negative impact on other forms of life. Further, nanoparticles can also be used as vectors for transporting the designed microorganisms or microbial consortia to the contaminated sites improving their survival, colonization ability and activity in otherwise unfavourable conditions.

Hence, nanoparticles present new approaches enabling microorganisms to break down micro/nanoplastics to effectively resolve the problems facing plastic pollution in matrices. Thus, nanoparticles serve as stabilizers of enzyme activity, increasing bioavailability, environmental cleaners, and models of controlled release systems—all necessary for implementing efficient models accelerating the degradation of polymeric material and decreasing its impact on the environment. Future work on nanoparticle-based strategies for effectively removing plastic pollution with minimal impact to the environment as well as human health will expand.

5.2 *Nano Bioengineered Compounds Improve Bioremediation Activities*

Leaking and channelling together with irresponsible disposal, dumping and burning of plastics are not sustainable, environmentally friendly strategies. Since it is a well-known fact that plastic wastes are polluting the environment to the maximum level, bioremediation is said to be a better solution to this problem. Over the last few years, self-assembled bioengineered nanomaterials have become important nanotools to improve bioremediation efficiency by providing better control over the degradation of plastic materials. Here, in this concluding section, the design aspects, uses, and probable advantages of the developed nanomaterials in the field of bioremediation of plastic have been detailed.

Bioengineered nanomaterials are nanostructures made from the interaction of fabricated biological elements with nanoparticles to provide certain characteristics to its bioremediation use. Several key design principles guide the development of biohybrid systems to enhance the bioremediation process of plastic waste:

(a) **Functionalization and surface modification**: The ability to coat nanoparticles with biomolecules like peptides, proteins or DNA/ RNA oligonucleotides put in a handler control on the interaction with plastic waste and other biological entities. Nanoparticles can undergo functionalization to enable them to adhere to various types of plastics hence enhancing their susceptibility to microbial degradation mechanisms.
(b) **Enzyme immobilization**: Their application is based on the possibility of using nanoparticles as encapsulating matrices for the enzymes participating in the paths of plastics degradation, such as lipase, esterase or oxidoreductase. It provides the immobilized enzyme with its catalytic properties while, in turn, offering a long-term and specific process of degrading plastic polymers into more manageable compounds for decomposing microorganisms.
(c) **Microbial growth**: Microbial cells or colonies are deployed to plastic immersed environments through bioengineered nanomaterials in a bid to enhance their colonization, activity and survival. Altered nanoparticles attract microbes to attach and form biofilms enhancing the process of plastic degradation.
(d) **Smart delivery systems**: Smart materials called nanomaterials can modify the delivery of bioactive agents including enzymes, nutrients or signalling molecules by detecting specific stimuli from the environment.

Bioengineered nanomaterials play vital role with variety of usages in the bioremediation of waste plastic, the multifaceted issues of plastic pollution are the given and sufficing solution to these issues is the focus of the present paper. These applications are based on the specific characteristics of nanoparticles and other artificial biological systems to enhance the rate, selectivity, and eco friendliness of the plastic waste disposal. Here are some of the major applications of bioengineered nanomaterials in the bioremediation of plastic waste:

(a) **Enhanced Plastic Degradation**: Bioengineered nanomaterials can improve degradation of plastic polymers via increased microbial enzyme interaction and metabolic process. Thus, nanoparticles could be used as carriers for the enzymes involved in the pathways of plastic degradation that include lipase, esterase or oxidoreductase. The immobilized enzyme still has its catalytic functionality and is stable and capable of continuously breaking down large plastic polymers selectively into smaller fragments which can be degraded by microorganisms.
(b) **Selective plastic targeting**: Selective functional nanoparticles are capable to recover specific types of plastics such as polyethylene, polypropylene or polystyrene. Coatings on nanoparticles include surface modifications by which the plastic matrix can be bound and specifically recognized to improve the effectiveness of the bioremediation process. Surface engineering and functionalization of developed bioengineered nanomaterials increases its selectivity and sensitivity towards target plastics hence reducing interference with non-target plastics.
(c) **Microbial metabolic engineering**: Self-assembling nanocarriers enable introduction of target genetic constructs (plasmids, DNA/RNA oligonucleotides or CRISPR-Cas systems) to microbial populations for metabolic reprogramming and control. It is a modality that permits microbial strains to degrade kinds of plastic polymers and adapt more effectively to environmental stressors in presence of plastics. The type of plastics that are targeted can be altered or the microbial communities which is created can be enhanced to work in a synergistic manner.
(d) **Smart delivery systems**: Smart nanomaterials that have some specific responsiveness can adjust the delivery of active substances—enzymes, nutrients, or signalling molecules-depending on the external stimuli. Self-assembling nanocomposites enhance the utilization of resources, reduce side reactions and overall improve the effectiveness of biodegradation processes. Encapsulation effectively permits precise control over the enzyme delivery and the rate at which the plastic is degraded without harming other species of life.
(e) **Biofilm formation and microbial adhesion**: The modified nanoparticles enhance microbial attachment and film formation on plastic substrates so as to enhance the plastics degradation process. Amphiphilic nanoparticles support microbial attachment and growth of plastic-degrading microbes on the surface of the F and P matrices. The current paper defines microbes' biofilms as localized substrate focused niches that can enhance enzyme activity and rate of degradation of plastics.

The bioengineered nanomaterials can be used to incorporate plastic waste into the environmental sensors to monitor the rate of bioremediation. Smart sensors may measure alterations to relevant environmental parameters including yours of degradation of plastics, microbial metabolism or even nutrient access to guide bioremediation and evaluate remediation. Nanoparticle based sensors are Mel, specific, sensitive, and have multiplex capability to track plastic dumpsites and active cleanup areas in a continuous manner. In general, the bioengineered nanomaterials have the capability to bring drastic change in the field of bioremediation of plastic waste by giving new solutions which will help in minimizing the issue of plastic waste in the environment.

6 Challenges and Ethical Considerations

The biodegradation of micro- and nanoplastic waste with the help of nanobiotechnologies today represents a significant number of questions and concerns that must be considered for the further effective and ethical use of these technologies. Below, we outline some of the main challenges and ethical issues in the use of nanobiotechnologies in this context.

6.1 Challenges

(1) **Environmental fate and toxicity**—Nanoparticles in processes related to biodegradation may present with unstable environmental characteristics and toxicity impacts. During biodegradation of polymer nanoparticles may dissolve into the environment and concentrate in the water and the soil affecting aquatic and terrestrial life forms. Long-term behavior, mobility and joking potential of ENPs in the bulk phase of environmentally relevant matrices is the major concern for risk assessment of ENPs to protect the ecosystem.

(2) **Efficacy and selectivity**—Selectively and effectively degrading micro and nano plastic waste using nanobiotechnology is one of the major technical challenges. Nanoparticles that have been fabricated intentionally require to be specific in the selective advection and depolymerization of plastic litter without causing havoc on other non-affected compounds and species. Enhancing the efficiency of nanobiotechnological techniques in the context of plastic depolymerization and selectivity necessitates the development of more complex nanomaterial design concepts and an understanding of the degradation kinetics and factors' influence.

(3) **Regulatory oversight**—Nanobiotechnology for biodegradation of plastic waste is regulated because engineered nanoparticles possess peculiar physicochemical properties and behavior. Modern regulatory legislation may have no clear criteria for evaluating the safety, efficiency and environmental friendliness of

nanobiotechnology products. It is important to establish check-and-balance patterns for the assessment and accreditation of nanobiotechnology solutions to eliminate the biodegradation of plastic waste to retain applicable responsibility and regulatory obscurity.

(4) **Scaling and deployment**—While nanobiotechnological processes of breaking down and degrading plastics from the laboratory scale to full-scale operation can present major engineering and logistical questions. To cultivate the engineered nanoparticles for wide use and to integrate those particles with the current waste disposal framework, the three key challenges of scalability, cost and environmental impact must be taken into consideration.

(5) **Equity and justice**—The equity in sharing benefits and risks from nanobiotechnological interventions in the biodegradation of plastic waste presents ethical issues (Ahmadi & Ahmadi, 2013). Making nanobiotech solutions for biodegradation of plastics inclusive for people in the lower end of the social scale is important due to what can be termed as environmental justice. Therefore, the lack of equity in availability of the necessary information, resources, and decision-making associated with nanobiotechnological interventions regarding plastic waste management deserves attention.

(6) **Risk perception and communication**—Therefore, it is pertinent to fully communicate and explain likely risks and uncertainties of nanobiotechnological interventions in the biodegradation of plastic wastes to nurture public acceptance. Stakeholders' awareness is enhanced through transparent dissemination of the scientific information, risk evaluations as well as regulatory measures. Nanobiotechnology in the management of plastic wastes is best deployed where the public has been informed and has removed misconceptions about nanobiotechnology displaced in a Lusaka market.

(7) **Environmental management**—Cultural and ethical issues concerning environmental management call for the use of precautionary and sustainable principles in the deployment and implementation of nanobiotechnological solutions to plastic wastes containments (Grunwald, 2020). Three major ethical principles that inform the development, utilisation and assessment of nanobiotechnology interventions regarding environmental stewardship and conservation include reduction of environmental risk, conservation of natural resources and conservation of biological diversity respectively. It is important to note that to encourage ethical conduct and effective regulation of plastic wastes while maintaining the ecosystems' integrity and sustainability a proper approach to sources' identification and disposal requires a system that supports integrated and ecosystem-based approach to the overall management of plastic wastes.

(8) **Public Participation**—Since ethical issues involving nanobiotechnological intervention for biodegradation of plastic wastes affect the stakeholders directly and indirectly, they need to be involved efficiently in any related project. Engaging stakeholders in the decision-making process, getting their input, and defining and endorsing the relevant values and norms regarding nanobiotechnology can enhance legitimacy through the processes of design and implementation. Engaging communities and building reciprocal relationships, respect as

well as trusting them to co-govern interventions in nanobiotechnology is vital to foster both ethical practices as well as corporate responsibility.

7 Future Prospects and Directions

Nanobiotechnology offers a promising opportunity to address the complex challenges posed by microplastic and nanoplastic pollution. Going forward, emerging trends, integration with circular economy principles, and collaborative efforts will play a key role in shaping the direction of research, innovation, and implementation of nanobiotech solutions to control plastic pollution.

7.1 New Trends in Nanobiotechnology in Micro/Nano Plastic Pollution Control

Nanoscale characterization techniques: Advances in nanoscale characterization techniques such as high-resolution microscopy, spectroscopy and imaging modalities will allow us to gain a deeper understanding of the interactions between nanoparticles, microorganisms and plastic contaminants. These techniques provide valuable insights into nanoscale plastic degradation mechanisms, nanoparticle behavior and microbial metabolism, informing the design and optimization of nanobiotechnology interventions.

1. **Biological nanoparticles** (Popescu et al., 2010): The use of biological nanoparticles synthesized from microorganisms, plants or marine organisms can provide a sustainable and environmentally friendly alternative to chemically synthesized nanoparticles. Bio-nanoparticles have unique properties and functions derived from natural biological processes, making them versatile and biocompatible for various applications in plastic pollution control (Mughal et al., 2021). The use of a variety of biological nanoparticles promises to develop innovative and environmentally sustainable nanobiotechnology solutions.
2. **Nanomaterials engineering**: Engineering nanomaterials with customized properties and functionality will drive innovation in plastic pollution control. Functional nanoparticles, nanocomposites, and hybrid materials can be engineered to generate specific interactions with plastic pollutants, microbial communities, or environmental matrices, improving their efficacy and selectivity in bioremediation processes. By leveraging advances in nanomaterial synthesis, surface modification, and self-assembly technologies, researchers can develop next-generation nanobiotechnology platforms for plastic waste management.
3. **Nanotoxicology and Risk Assessment**: As the use of nanobiotechnology is increasing, the importance of nanotoxicology and risk assessment cannot be underestimated. Understanding the potential environmental and health risks associated with engineered nanoparticles is critical to ensure safe and responsible

deployment of nanobiotechnology solutions for plastic pollution control. An integrated approach combining experimental studies, computational modelling, and predictive analysis will enable comprehensive risk assessment and informed decision making regarding the use of nanobiotechnology in environmental applications.

4. **Smart nanomaterials and sensing platforms**: The development of smart nanomaterials and sensing platforms will enable real-time monitoring and control of the plastic pollution remediation process. Responsive nanomaterials are able to sense environmental cues such as plastic concentrations, microbial activity or nutrient availability and can dynamically adjust their properties and functionality to optimize remediation efficiency. Integrated sensor networks and IoT-enabled platforms provide opportunities for remote monitoring, data analysis and decision support systems, facilitating adaptive management of plastic pollution hotspots and remediation sites.

7.2 Integration of Nanobiotechnology and Circular Economy Principles

The circular economy model provides a holistic approach to resource management that emphasizes reducing, reusing, recycling, and recovering materials to minimize waste generation and environmental impact. Combining nanobiotechnology with circular economy principles offers a transformational opportunity to address the challenge of plastic pollution while improving sustainability and resource efficiency (Zhu, 2022). Key aspects of this integration include:

1. **Resource recovery and reuse**: Nanobiotechnology can develop innovative resource recovery and reuse strategies for plastic waste. Engineered nanoparticles, biodegradable polymers, and bio-based materials can be used to convert plastic waste into value-added products such as biodegradable plastics, biofuels, or biochemicals. Using enzymatic degradation, microbial fermentation, or bioconversion processes, nanobiotechnology approaches can promote the circularity of plastic waste streams, reduce dependence on virgin resources, and minimize environmental footprints (Chakraborty et al., 2023).
2. **Closed-loop supply chain**: Implementing a closed-loop supply chain for plastics involves designing products, materials, and processes while taking end-of-life issues into account. Nanobiotechnology plays a key role in designing biodegradable, recyclable, and compostable materials that can be reincorporated into the production cycle. By optimizing physical properties, degradation kinetics, and environmental compatibility, nanobiotechnology interventions promote a closed-loop system where plastic waste can be continuously recovered, recycled, and reused, reducing reliance on linear single-use plastics.
3. **Life cycle assessment (LCA) and eco-design**: Integrating lifecycle assessment (LCA) principles and eco-design methods into nanobiotechnology innovation can promote environmental sustainability and circularity. Life cycle

assessment enables a comprehensive evaluation of the environmental impact of nanobiotechnology interventions throughout the entire product life cycle, from raw material extraction to end-of-life disposal (Kratošová et al., 2019). The eco-design approach focuses on optimizing the environmental characteristics of nanobiotechnology products and processes, taking into account factors such as material selection, energy consumption, and waste generation. By adopting a life cycle perspective and incorporating eco-design principles into nanobiotechnology development, stakeholders can identify opportunities to optimize resources and reduce waste.

4. **Policy and regulatory framework**: Policy and regulatory frameworks are crucial to support the integration of nanobiotechnology with circular economy principles to tackle plastic pollution. These frameworks provide the necessary guidelines incentives, and safeguards to promote sustainable practices, innovation, and responsible resource management (Venkatesh, 2022). Here is how policy and regulatory frameworks can promote the integration of nanobiotechnology with circular economy principles:

   (a) Encouraging innovation: Policy incentives (such as research funding, tax credits, and grants) can encourage investment in research and development of nanobiotechnology for plastic waste management. By providing financial support and incentives for innovative solutions in line with circular economy goals, governments can encourage the adoption and scale-up of nanobiotechnology interventions.

   (b) Regulatory compliance: Regulatory frameworks play a critical role in ensuring the safety, effectiveness and environmental sustainability of nanobiotechnology products and processes. Regulations regarding the use of nanoparticles, biodegradable polymers and bio-based materials in plastic waste management should be based on solid scientific evidence and risk assessment. Standardized protocols, testing procedures and monitoring requirements are critical to assess the safety and performance of nanobiotechnology interventions and to ensure compliance with regulatory standards.

   (c) Environmental impact assessment: Assessing the environmental impact of nanobiotechnology interventions is important to assess their sustainability and compatibility with circular economy principles. Environmental impact assessment (EIA) helps to identify potential risks, minimize adverse effects, and optimize the environmental performance of nanobiotechnology products and processes. By incorporating EIA into the regulatory approval process, policymakers can ensure that nanobiotechnology solutions for plastic pollution control meet environmental standards and contribute to overall sustainability goals.

   (d) Stakeholder engagement: Stakeholder engagement, including policymakers, industry representatives, researchers, and civil society organizations, is critical to developing effective policy and regulatory frameworks for the integration of nanobiotechnology. Stakeholder consultations, public hearings, and participatory decision-making processes promote transparency, accountability, and inclusiveness in policy development. By seeking input from diverse stakeholders, policymakers can address issues, build consensus, and generate support for nanobiotechnology solutions that promote circular economy principles.

(e) International Collaboration: Plastic pollution is a global challenge that requires coordinated action and collaboration across borders. International collaboration mechanisms such as bilateral agreements, multilateral initiatives and regulatory coordination efforts facilitate knowledge sharing, capacity building and technology transfer in the field of nanobiotechnology for plastic waste management.

By harmonizing regulatory approaches and sharing best practices, countries can harmonize standards, streamline approval processes and promote global adoption of nanobiotechnology solutions to control plastic pollution based on a circular economy. Policy and regulatory frameworks play a key role in supporting the integration of nanobiotechnology with circular economy principles for plastic pollution control. By encouraging innovation, ensuring regulatory compliance, conducting environmental impact assessments, engaging stakeholders, and promoting international collaboration, policymakers can promote a circular economy for plastics by creating an enabling environment for the development, deployment, and adoption of sustainable nanobiotechnology solutions.

8 Conclusions

Micro/nano plastic pollution represents a pressing environmental challenge with far-reaching consequences for ecosystems, biodiversity, and human health. This pervasive issue necessitates immediate, innovative, and multifaceted approaches to mitigate its impact. Among the available solutions, nanobiotechnology emerges as a powerful tool to address the problem at its root and across its lifecycle, offering groundbreaking capabilities for detection, degradation, recycling, and remediation.

The application of nanosensors for real-time monitoring provides critical insights into the distribution and fate of micro/nanoplastics, enabling precise and targeted interventions. Innovations such as enzyme-based degradation systems and nanoparticle-assisted technologies present promising pathways for breaking down persistent plastics into non-toxic byproducts. In the realm of waste management, nanobiotechnology enhances recycling processes and supports the transition to a circular economy through biodegradable nanomaterials and waste-to-energy solutions. Furthermore, nanoparticle-based soil amendments, filtration systems, and engineered microorganisms are reshaping environmental remediation efforts by effectively removing micro/nanoplastics from contaminated ecosystems.

To fully realize the potential of nanobiotechnology, sustained investment in research and development is paramount. This includes advancing nanosensor technologies to improve their sensitivity, selectivity, and scalability, as well as optimizing nanomaterial-based degradation processes to ensure their long-term environmental safety. Interdisciplinary collaboration among scientists, policymakers, and industry stakeholders is essential to foster innovation, address technical and ethical challenges, and translate laboratory breakthroughs into real-world applications.

Policy measures and regulatory frameworks play a crucial role in preventing further plastic pollution. Initiatives such as extended producer responsibility, ecodesign principles, and sustainable consumption practices can significantly reduce the influx of plastics into the environment. Equally important is raising public awareness through education campaigns to encourage individuals to adopt eco-friendly habits and advocate for collective action.

In conclusion, combating micro/nanoplastic pollution demands a unified global effort, leveraging the cutting-edge capabilities of nanobiotechnology alongside policy innovation and societal engagement. By integrating these approaches, we can pave the way for a cleaner, healthier, and more sustainable future. As we confront this urgent environmental challenge, it is imperative to act decisively, promote collaboration, and commit to safeguarding our planet for generations to come.

References

Acarer, S. (2023). Microplastics in wastewater treatment plants: Sources, properties, removal efficiency, removal mechanisms, and interactions with pollutants. *Water Science & Technology, 87*(3), 685–710.

Ahmadi, M., & Ahmadi, L. (2013). Ethical considerations of nanobiotechnology. *Journal of Biomaterials and Tissue Engineering, 3*(3), 335–352.

Allouzi, M. M. A., Tang, D. Y. Y., Chew, K. W., Rinklebe, J., Bolan, N., Allouzi, S. M. A., & Show, P. L. (2021). Micro (nano) plastic pollution: The ecological influence on soil-plant system and human health. *Science of the Total Environment, 788*, 147815.

Bhattacharjee, J., & Roy, S. (2023). Green remediation of microplastics using bionanomaterials. In: *Remediation of plastic and microplastic waste* (pp. 240–260). CRC Press.

Bhoi, N. K., Singh, H., & Pratap, S. (2020). Developments in the aluminum metal matrix composites reinforced by micro/nano particles—A review. *Journal of Composite Materials, 54*(6), 813–833.

Chakraborty, D., Chatterjee, S., Althuri, A., Palani, S. G., & Mohan, S. V. (2023). Sustainable enzymatic treatment of organic waste in a framework of circular economy. *Bioresource Technology, 370*, 128487.

Claessens, M., Van Cauwenberghe, L., Vandegehuchte, M. B., & Janssen, C. R. (2013). New techniques for the detection of microplastics in sediments and field collected organisms. *Marine Pollution Bulletin, 70*(1–2), 227–233.

David, M., Prakash, L., Sangeetha, J., Naik, J., Thangadurai, D., & Thimmappa, S. C. (2020). Degradation of plastics using nanomaterials. In *Handbook of nanomaterials and nanocomposites for energy and environmental applications* (pp. 1–13).

De, B., & Goswami, T. K. (2022). Nanobiotechnology—A green solution. In *Biotechnology for zero waste: Emerging waste management techniques* (pp. 379–396).

Gagliardi, M., & Mazzolai, B. (2015). Molecularly imprinted polymeric micro-and nano-particles for the targeted delivery of active molecules. *Future Medicinal Chemistry, 7*(2), 123–138.

Grunwald, A. (2020). Nanotechnology—A new field of ethical inquiry? In *The ethics of nanotechnology, geoengineering, and clean energy* (pp. 17–31). Routledge.

Haupt, K., Linares, A. V., Bompart, M., & Bui, B. T. S. (2011). Molecularly imprinted polymers. In *Molecular imprinting* (pp. 1–28).

Kamath, S. V., Mruthunjayappa, M. H., Mondal, D., & Kotrappanavar, N. S. (2022). Nanocomposite-based high-performance adsorptive water filters: Recent advances, limitations, nanotoxicity and environmental implications. *Environmental Science: Nano, 9*(7), 2320–2341.

Kathiriya, T., Gupta, A., & Singh, N. K. (2021). An opinion review on sampling strategies, enumeration techniques, and critical environmental factors for bioaerosols: An emerging sustainability indicator for society and cities. *Environmental Technology & Innovation, 21*, 101287.

Khan, M. S., Asif, M. I., Asif, M., Khan, M. R., Mustafa, G., & Adeel, M. (2024). Nanomaterials for the catalytic degradation and detection of microplastics: A review. *Topics in Catalysis*, 1–18.

Kratošová, G., Holišová, V., Konvičková, Z., Ingle, A. P., Gaikwad, S., Škrlová, K., Prokop, A., Rai, M., & Plachá, D. (2019). From biotechnology principles to functional and low-cost metallic bionanocatalysts. *Biotechnology Advances, 37*(1), 154–176.

Mohan, B., Singh, K., Gupta, R. K., Kumar, A., Pombeiro, A. J., & Ren, P. (2024). Water purification advances with metal–organic framework-based materials for micro/nanoplastic removal. *Separation and Purification Technology*, 126987.

Mughal, B., Zaidi, S. Z. J., Zhang, X., & Hassan, S. U. (2021). Biogenic nanoparticles: Synthesis, characterisation and applications. *Applied Sciences, 11*(6), 2598.

Musa, I. O., Auta, H. S., Ilyasu, U. S., Aransiola, S. A., Makun, H. A., Adabara, N. U., Abioye, O. P., Aziz, A., Jayanthi, B., Maddela, N. R., & Prasad, R. (2024). Micro- and nanoplastics in environment: Degradation, detection, and ecological impact. *International Journal of Environmental Research, 18*(1), 1.

Nzeyimana, B. S., Chakraborty, S., Priyadharshini, R., Mary, M. S., & Govindaraju, M. (2025). Micro/nano-plastic pollution in aquarium systems: Sources, fate, hazards, and ecological imbalances. In Global impacts of micro- and nano-plastic pollution (pp. 225–256). IGI Global.

Padervand, M., Lichtfouse, E., Robert, D., & Wang, C. (2020). Removal of microplastics from the environment. A review. *Environmental Chemistry Letters, 18*(3), 807–828.

Pelegrini, K., Pereira, T. C. B., Maraschin, T. G., Teodoro, L. D. S., Basso, N. R. D. S., De Galland, G. L. B., Ligabue, R. A., & Bogo, M. R. (2023). Micro- and nanoplastic toxicity: A review on size, type, source, and test-organism implications. *Science of the Total Environment, 878*, 162954.

Peng, L., Fu, D., Qi, H., Lan, C. Q., Yu, H., & Ge, C. (2020). Micro- and nano-plastics in marine environment: Source, distribution and threats—A review. *Science of the Total Environment, 698*, 134254.

Popescu, M., Velea, A., & Lőrinczi, A. (2010). Biogenic production of nanoparticles. *Digest Journal of Nanomaterials & Biostructures (DJNB), 5*(4).

Reddy, A. B., Reddy, G. S. M., Sivanjineyulu, V., Jayaramudu, J., Varaprasad, K., & Sadiku, E. R. (2016). Hydrophobic/hydrophilic nanostructured polymer blends. In *Design and applications of nanostructured polymer blends and nanocomposite systems* (pp. 385–411). William Andrew Publishing.

Varshney, A. (2024). Cleaning up the smallest pollutants: The potential of microbial degradation in tackling micro- and nano-plastic pollution. In *Management of micro and nano-plastics in soil and biosolids: Fate, occurrence, monitoring, and remedies* (pp. 367–389). Springer Nature Switzerland.

Venkatesh, G. (2022). Circular bio-economy—paradigm for the future: Systematic review of scientific journal publications from 2015 to 2021. *Circular Economy and Sustainability, 2*(1), 231–279.

Youssef, A. M., Malhat, F. M., & Abd El-Hakim, A. F. A. (2013). Preparation and utilization of polystyrene nanocomposites based on TiO_2 nanowires. *Polymer-Plastics Technology and Engineering, 52*(3), 228–235.

Zhu, H. (2022). Research on promoting the development of circular economy by using nano-biopharmaceutical technology. *International Journal of Nanotechnology, 19*(6–11), 959–969.

Assessing Effects and Impacts Associated with Macro–Micro Plastics in Waterbodies

T. Rajendra Prasad, K. Rama Krishna, K. V. Sharma, and C. Nagabhaskar

Abstract The amount of plastic produced worldwide has increased to billions of tons every year, raising serious concerns about the environment and human health. There are currently no industrial or naturally occurring processes that can fully break down plastics into innocuous leftovers. Nearly 80% of the planet's water, which makes up two-thirds of it, is contaminated by plastic waste. Plastic-induced water pollution is putting marine life in dangerous and terrible conditions. The primary culprits behind water body pollution are micro- and macroplastics, which not only cause various health problems but also directly endanger the natural habitation of freshwater and sea plants and animals. Recent population growth, industrial burst, and human activity have increased the need for plastics in the production of various goods, the majority of which pollute freshwater and sea environments. We have examined the three domains of solutions that can be used to combat plastic pollution and remediate plastic-induced water pollution. These include the use of cutting-edge technology to create bioplastics, the use of microorganisms to biodegrade plastics, and raising public awareness and enacting laws. Universities, research institutions, nonprofit groups, and international agencies may raise alertness in our people about the need to fight water contamination by assisting governments worldwide in developing and enforcing new laws that effectively battle plastic-induced water contamination.

Keywords Plastic pollution · Macroplastics · Microplastics · Freshwater organisms · Marine organisms

---

T. Rajendra Prasad (✉) · C. Nagabhaskar
NRI Institute of Technology, Agiripalli, Vijayawada, India
e-mail: trajendramech@gmail.com

K. Rama Krishna
Department of Mechanical Engineering, Koneru Lakshmaiah Education Foundation, Guntur, India

K. V. Sharma
Centre for Energy Studies, JNTUH College of Engineering, Hyderabad, India

P. Pathak et al. (eds.), *Plastic Footprint*, https://doi.org/10.1007/978-981-96-6362-0_7

1 Introduction

Plastics are chemical compounds that are produced or obtained from sources such as coal, natural gas, and crude oil through polymerization method (Dang et al., 2018). Our knowledge of polymer technology was completely transformed in 1907 when Bakelite, the first man-made plastic, was discovered. This was because it sparked the creation of several polymers, plastic formulas, and development methodologies (Frias & Nash, 2019). We encounter plastic on a daily basis more than any other particular substance because it is the most widely used product in our homes, businesses, and technological systems. Approximately 140 million tons of plastic are produced annually by global industry, with a larger percentage of that quantity ending up in the environment as trash, according to current statistics. Polyethylene, which has a high molecular mass and hydrophobicity, makes up around 64% of man-made plastics. A significant portion of the 1 trillion polythene products produced globally each year wind up in the land and water. Materials used for packaging, such as polythene, are responsible for about 10% of the urban waste generated globally. But only 5% of the waste is recycled; the rest is buried underground, where it takes about a century for the material to break down physically without the help of microorganisms. This is the outcome of their tenacity, resilience, and incapacity to deteriorate. A polymer's susceptibility to degradation is significantly influenced by its chemical makeup, which includes its arrangement, the presence of certain atoms, and other chemicals. The inertness of polymers with simple carbon chains, especially those with double bonds, is higher than that of polymers which consist of heteroatoms or additions. Because of their high degree of purity, they are less vulnerable to outside effects, which slow down the break down process (Agarwal et al., 2024; Muneer et al., 2021). Once considered a great triumph, the discovery of plastic has now descended into an environmental catastrophe, polluting not only our land but also our surface, aquifers, and marine water sources. Building, electronics, automotive, and wrapping are the main industries that use over 70% of the plastic generated globally. Packaging accounts for one-third of the yearly demand for plastic in Europe alone. Less than 10% of plastic gets recycled; the remainder is burned or dumped in landfills (Thevenon & Carroll, 2015). According to current research undertaken worldwide, plastics are the most serious and significant contributors to environmental degradation.

Water is a widespread solvent that is necessary for practically all chemical and natural processes that occur within the bodies of living things. Water makes up as much as 60% of the grown-up human being. We are unable to carry on long times lacking water, yet we are able to survive longer times without nutrition. Water accessibility is essential for natural species to survive in the biosphere. Economic expansion depends on agricultural and industrial advancement, but inadequate water supply, water shortage, or contaminated water can cause major economic troubles like low infrastructure growth, low per capita earnings, and other issues that impede and impact a state's gross domestic product due to water contamination alone (Lamb et al., 2018). According to reliable estimates, water covers two-thirds of the Earth's surface, with the remaining portion being land. 98% of the world's water is found in the oceans and can't be utilized for intake lacking treatment, such as filtration, which undoubtedly charge a lot to meet the population's increasing requirement. Up to 40% of our oceans are severely impacted by human activity today. Approximately 2% of all the water on globe is freshwater. Nevertheless, only 0.4% of this water is currently available for human consumption; the remaining 1.6% is frozen into glaciers and ice caps, which prevent us from using them. If we do not protect these water assets, the future seems bleak. Up to 40% of our marine waters are severely impacted by human actions like plastic waste today. The amount of plastic-related pollution in marine waters is at an all-time high and is getting worse every day. In the upcoming years, the issue will only get worse if immediate action is not taken. Numerous aquatic animals are already suffering from plastic trash, and it's likely that it will also comprise an effect on people's well-being in ways that we do not so far completely recognize (Villarrubia-Gómez & Fabres, 2018). The existence of plastics in the remains of marine birds and animals was the subject of the first technical papers addressing the plastic contamination in our marine sources, which were published in 1969 (Pietrelli et al., 2018). This pollution may put humans at serious risk for biological and healthiness problems. Microplastics are especially concerning since they have been connected to immediate death in sea animals (Agarwal et al., 2024). In addition to researching the reasons and detrimental impacts of plastic water effluence on ecological and mankind issues, our work focuses on finding sustainable ways to reduce plastic water pollution (Barboza et al., 2018; Mrowiec, 2017).

2 Water Pollution and Causes

2.1 Water Pollution

The accumulation or overabundance of one or more pollutants in water can cause problems for aquatic life, plants, animals, and people. Water contamination is the term for this. Plastics have been found in marine environments all around the world. Numerous studies have calculated that over 5 trillion parts of plastic waste remain afloat in the seas. Nearly 80% of the ocean trash, according to the United Nations, is plastic or produced from plastics, and it affects about 800 species worldwide. Nearly 67% of sea plastic is estimated to originate from land-based causes, with the remainder originating from ocean-based causes (Karbalaei et al., 2018; Kushwaha et al., 2024). An estimated 130 lakh tons of plastic debris wind up into the seas annually. Plastic waste can be consumed by various marine creatures leading to their eventual asphyxia, malnutrition, and drowning. According to scientific estimates, around 60% of all seabirds have consumed plastic bits at some time in their lives. The study in question raises concerns since it predicts that by 2050, this percentage would rise to 99% (Sharma & Chatterjee, 2017). States all across the world are enforcing stringent laws to lessen the negative effects of plastic pollution. To achieve this, states have outlawed one-time-use plastics, replaced crude-oil-based plastics with bio-friendly alternatives like glass, paper, or eco-friendly plastics, and improved squander gathering methods to ensure that all squander is appropriately gathered, recycled, or disposed of. Guidelines on steps that governments, businesses, and public sector organizations can take to reduce the manufacture and use of single-use plastics are provided by the United Nations' program on Environment. These directives and recommendations include enhancing waste managing schemes; charitable turn down agreements and initiatives; raising public alertness and force; and encouragement of environmentally sustainable alternatives (Godfrey, 2019).

By 2050, plastics in the sea count more than fish, according to the Ocean Unite website, which estimates that 80 lakh tons of plastics go into sea yearly. Over the past 50 years, the rate of plastics manufacturing worldwide has been steadily rising at a rate that is significantly faster than anticipated. The quantity of plastic waste is increasing annually despite the latest economic predicament. In 2010 alone, between 50 and 130 lakh tons of plastic debris reached the sea from land sources (Geyer et al., 2017). Plastics are thought to be produced using 4–5% of the world's yearly petroleum production, while the same quantity of energy resources, such as electricity and petroleum, are used for how these plastics are made. Both point and nonpoint sources can contribute to contamination. Point caused contamination is the word used to describe pollution that come from a lone cause. For instance, an industry, a refinery, or sewage water remedial facilities may with permission or unlawfully emit wastewater, also known as effluent. Rainstorm sewer releases and wastewater runoff, whether urban or industrial, are examples of point sources of contamination that primarily impact the area next to them. Although point source contamination originates in a lone site, it can boast an effect on miles of streams and the seas. Pollutants that originate from a variety of unidentified causes and enter aquifers or surface water through a multitude of pathways are considered nonpoint sources of contamination. These could be farming runoff, rainstorm water runoff, or trash that is carried into streams from the terrain. The primary and indisputable basis of water contamination in United States is nonpoint-based infectivity, which is tricky to manage because there isn't a lone, recognizable root (Sigler, 2014).

2.2 *Causes of Water Pollution*

Numerous factors, including raw industrial effluents, municipal garbage, and the leakage of heavy metallic species from top soil into aquifers, contribute to water contamination. Human activity is the primary cause of contamination in the sea and the majority of surface water systems. Products created as a result of urbanization have become pollutants because of their excessive usage, manufacture, and non-disposable circumstances. One of the main global contributors to water contamination is plastic. Since the topic of this chapter is plastic-induced water contamination, we will go into great depth regarding how plastics contaminate both fresh and seawater sources. There is also discussion on how to combat water contamination.

2.2.1 Plastic as a Water Pollutant

The Greek sound "plastikos," which means "capability to modify into diverse shapes," is where the word "plastic" originates (Shah et al., 2008). Analogous to naturally occurring biopolymers like wool or wood, which are massive molecules known as polymers made up of repeating units called monomers with carbon backbones, plastics are manufactured organic polymers (Halden, 2010). Petrochemicals and their derived yields are heated in regulated conditions to create polymers, which are then broken down into smaller molecules called monomers or simpler residues (Kibria et al., 2023). Different kinds of monomers result in plastic resins with varying properties, like strength or shaping ability. Coal, natural gas, or oil can all yield those monomers. Polypropylene, polyvinyl chloride, polystyrene, polyethylene, and polyethylene terephthalate are the five main kinds of plastics (which account for 90% of polymer making) that are more prevalent in the ocean (Thevenon & Carroll, 2015). However, because untainted plastics have poorer thermal and material qualities and increased deterioration, they require additives like stabilizing materials to achieve specific required attributes. In addition to making life easier for people, the growth of plastics produces a lot of waste, which endangers the environment and all living things. Because of this large quantity of desecrate plastics, the ecology is impacted by soil contamination from land-filling, marine contamination from ocean dumping, and air pollution from open dumping. Figure 1 offers more details on how this plastic dispel is anticipated to be handled in a scenario where production as usual is the norm. By 2060, plastic trash is expected to exceed 1 billion tons per year, as seen from Fig. 1. Waste in landfills continues to be the most common type. A significant risk to ecosystems is highlighted by the projection that unmanaged plastic squander, which comprise tormented or improperly disposed of plastics, will surpass two hundred tons by 2060. If significant steps are not done, the amount of plastics in the seas may eventually exceed the mass of all fish by 2050 if existing trends continue. The projected increase in the production of plastic garbage, together with the slow advancement of recycling and environmentally friendly disposal techniques, highlights the urgent need for revolutionary international action (Fayshal, 2024).

Additionally, these problems will compound over the course of plastic's existence sequence and pose a threat to ecological balance (Paletta et al., 2019). The size of plastic wreckage varies, further dividing them into three categories: macro-fragmented, measuring 5 mm; micro-fragmented, measuring 1 μm to 5 mm; and nano-fragmented, measuring less than 2 mm. Below is a quick discussion of each of these.

2.2.2 Macroplastics

The larger plastic particles, known as macroplastics, are usually bigger than 5 mm and are particularly prevalent in maritime environments. The length of time that macroplastics stay in the sea varies; some examples include fishing lines (600 years),

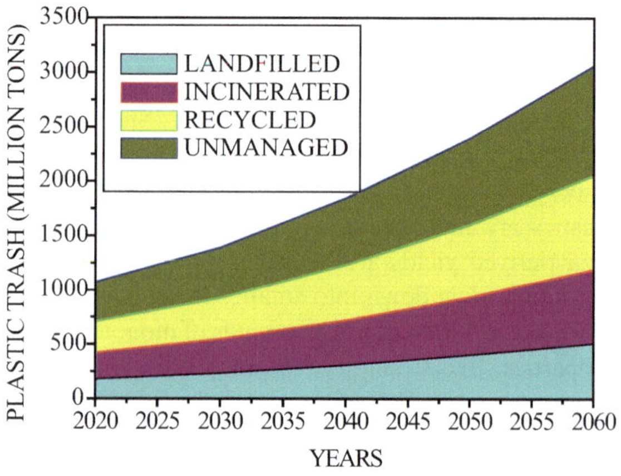

Fig. 1 Projected plastic waste by various disposal methods

plastic six-pack rings (400 years), plastic bags (50 years), polyethylene bottles (450 years), cigarette ends (1–5 years), and polystyrene cups (50 years).

2.2.3 Microplastics

Microplastics are solid units, just like the majority of conventional synthetic polymers. These synthetic polymers can vary in extent from 1 μm to 5 mm, and their shapes can be ordered or random. There exist two types of microplastics: primary and secondary, depending on where they come from (Ivar doSul & Costa, 2014). Primary microplastics are tiny particles, but secondary microplastics are made when larger particles crack down into smaller ones and generate plastic. Microplastics are plastic particles smaller than 5 mm in diameter, which may also comprise particles in the nano-size range, or 1 nm, according to the joint panel of experts on marine environmental protection (Beaumont et al., 2019). Both primary and secondary sources contribute to marine microplastic pollution (Andrady, 2011). Primary microplastics originate from a variety of sources, including everyday items like toothpaste, cosmetics, detergents, cleaning and maintenance supplies for business and industry, shower gels, shampoos, and laundry fibers. About 90% of microplastics can be stored in wastewater treatment facilities (Law, 2017). In a relatively short period of time, microplastics have greatly expanded into a variety of ecosystems, from earthly habitats to the depths of the vast ocean. Positive findings regarding the occurrence of microplastics have also been observed in the polar areas. Because microplastics are so tiny, many marine creatures are vulnerable to becoming entangled in them, which can lead to physical harm, asphyxia, and hunger. This raises serious environmental issues and increases the dangers to plants and animals. Larger plastic components break down to produce secondary microplastics. The majority of secondary microplastics are formed by the physical, chemical, and biological cracking of larger particles, which leads to their direct release into marine habitats (Wang et al., 2018).

2.2.4 Nanoplastics

Particles between 1 nm and 1 μm in size that are created when plastic things degrade are known as nanoplastics (Frias & Nash, 2019). The breakdown of plastic doesn't end at the micro-level; rather, it keeps getting smaller and smaller until it reaches nanoscale plastic particles. Nanoplastics are created by further breaking down nearly all microplastics from laundry or microbeads found in cosmetics. There is a dearth of scientific research and studies on nanoplastics, and it is still unknown how harmful they are to aquatic life. Numerous systematic studies conducted in the ocean have demonstrated that plastic particles can be consumed by a variety of aquatic organisms, such as fish and tortoises, and that they can travel from the intestines into the animal's circulatory system, ultimately triggering an immune response (Amelia et al., 2021; Ivanova et al., 2018). It is possible for ingestion of these nanoparticles to result in bioaccumulation. The food chain may be disrupted as a result of the bioaccumulation of plastic particles in various creatures that live in a variety of water bodies. Typically, bioaccumulation begins with simpler microbes and progresses to higher animals in the food chain, such as fish after algae and zooplankton. According to research, these tiny plastic elements that build up within the bodies of various organisms have the ability to enter the brain and cause behavioral problems. Nevertheless, lots of animals in nature are infrequently exposed to nanoparticles over their lifetimes (Rhodes et al., 2018).

3 Plastics and Marine Debris as Pollutants

Waste made of plastic is a persistent pollutant or contamination that does not break down in the environment. It readily clings to water-repellent organic contaminants, which are enduring and strongly linked to the death of numerous aquatic species (Barboza et al., 2018). Their overabundance in contemporary civilization is demonstrated by the enormous mountains of waste that surround us. Glass, metal particles, paper, textile products, wood, rubber, and plastics are examples of solid materials that are discharged into aquatic environments and are referred to as marine or aquatic waste. While the majority of these materials are continual and long-standing in the marine atmosphere, others are biodegradable. The consequences of sunken wooden vessels and their ceramic relics in water bodies are minimal because they are not constructed of plastic (Ahmed et al., 2018). Contrarily, plastics are distinct in that they are resistant to biodegradation and, because of their low weight, are easily transported by wind and water (Law, 2017). Globally, plastics production has surged. Aquatic life is seriously threatened by the occurrence and hazards of plastics in the marine environment, which motivates new study to assess plastic pollution and its effects on the economy, ecology, biology, and society as well as to uncover potential solutions. There is a lot of plastic trash in coastal seas and urban beaches. Every surroundings, including freshwaters, beaches, the deep seafloor, and the water column, has microplastics. Plastic's negative effects on the surroundings include harm to aquatic life, health hazards, and habitat (Bråte et al., 2017).

3.1 Microplastics and Microorganisms

Microbes like bacteria and archaea are naturally occurring recyclers and biodegraders; they may also help break down plastics' chemical leftovers (Yu et al., 2018). By transforming the carbon that makes up plastics into carbon dioxide (CO_2), water, and biomass, microorganisms can decompose polymers. When microplastics float on water, they can serve as raft substrates for different microorganisms and epifauna, such as bacteria, algae, and diatoms, and move them to areas where they were previously absent. This can alter the marine ecosystem by upsetting the microbes' natural habitat, which ultimately affects the food chain (Mrowiec, 2017).

4 Plastic Pollution in Rivers

Plastic pollution is a growing environmental hazard that has an adverse effect on rivers, lakes, and oceans worldwide. It is believed that rivers serve as the primary routes for plastic debris from the land to reach the ocean (Strokal et al., 2023). Only in the early 2010s were the first attempts to measure the flow of riverine plastic trash were out, and they involved sampling streams in North America and Europe, including the Tamar Estuary, the Danube, the Seine, the Los Angeles region, and the Thames River. The field-based studies provide crucial information about the types of plastic polymers, the amounts of plastic waste, and how they change over time and space. The first worldwide estimates of plastic emissions into the oceans were supplied using modeling methodologies, which complemented case studies unique to individual rivers. Because they provide a first-order prediction of the anticipated size and geographic distribution of riverine plastic transport, these assessments are incredibly helpful. By using these insights, geographical knowledge gaps may be identified, and plastic trash measurement techniques can be improved and harmonized. The majority of poorly managed plastic garbage, according to mounting observational data, ends up in plants, estuaries, riverbanks, and on land. Years, decades, or even longer may be spent here. River features and hydrological conditions are the primary drivers of river plastic transit and retention. River discharge is frequently linked to the seasonal fluctuation in floating plastic transport, with higher discharge typically resulting in higher transport rates. The dynamics of plastic's transit and retention in rivers have been thought to be significantly influenced by river discharge peaks in particular. Optimizing prevention and reduction efforts requires an understanding of how plastic transport and retention react to extreme discharge conditions, such as floods (Pinto et al., 2024).

The dynamics of plastic pollution's transit and retention are significantly influenced by floods. Increased water level and flow velocity mobilize more plastics, which can either enter the aquatic environment directly or travel through urban drainage and sewer systems. Once the water level and flow velocity are sufficiently enough, accumulated plastics in sediment, on riverbanks, or in floodplains may also be remobilized. According to estimates, floods with a 10 year recovery period already have the capacity to mobilize ten times as much plastic garbage globally as non-flood conditions. Floods' impact on river plastic contamination is not clear-cut. More plastics are washed into the river by floods, but rising water levels can also trap more plastics in riparian plants and cause deposition on floodplains. Therefore, floods serve as both a transport and accumulation driver. The impact of floods on the movement and buildup of plastic is still unclear because there are few observations at the watershed scale (Strokal et al., 2023; Pinto et al., 2024; Emmerik & Schwarz, 2020).

5 Plastic Pollution in Oceans

The majority of plastic contamination is found along the coasts of wealthy nations like the US and China. The improper handling of about one-third of the plastic garbage makes it possible for it to enter the oceans. Roughly 90% of this plastic pollution is thought to have originated from just ten rivers worldwide, six of which are in Asia and two in Africa (Ragossnig & Agamuthu, 2021). Textiles (13.9% of all plastic trash), wrapping (46.7%), and consumer-institutional products (12.3%) are the three primary industrial areas that produce the most plastic garbage. Our environment and oceans are being severely contaminated by the estimated 0.11 million tons of pelagic plastic dissipate that originate in the United States annually (Wang et al., 2018). Although plastic from different sources enters the oceans in different ways, it is yet uncertain which diversified sources contribute to ocean contamination. There is now a substantial knowledge gap since reliable quantitative comparisons between plastic contribution masses, bases, and originating industries cannot be produced. Among land-based origins, rivers, wastewater pipelines, and shorelines are the main ways that plastics end up in the ocean. Large volumes of plastic remains can enter the oceans as a result of waste management practices and specific meteorological events like storms and strong winds, which are challenging to address. Ocean-based sources at sea can include intentional disposal of plastic debris in sea- or ocean-based travels, shipping operations including cargo ships that convey products, oil tankers, and other sea liners or vessels, or regular unintentional losses (Bråte et al., 2017). The land-based sources of macroplastics include construction, food, packaging, travel, household products, and drink containers. A significant portion of the aquatic plastic pollution that ends up in the ocean and sea comes from land-based sources, amounting to between 4.8 and 13 million metric tons annually. The shipping industry and fishing operations are two marine-based sources of macroplastics. Furthermore, the majority of aquaculture structures are composed of plastic, and if they are not maintained or are harmed by the environment, they may cause water to become contaminated with plastic (Haward, 2018).

6 Plastic Pollution in Groundwater

Concerns regarding aquifer preservation and contamination sources, particularly the so-called "emerging contaminants," may arise from an increase in human pressure. Microplastics (MPs) are a relatively new worry regarding plastic pollution, despite the fact that the effects of macroplastics, such as plastic baggage and containers, have been thoroughly researched and documented, particularly in marine settings. MPs, which are microscopic plastic elements less than 5 mm, are now known to be widespread and persistent contaminants in a variety of environmental niches, such as freshwater and marine ecosystems. Concerns over MPs' possible effects on the environment and human health are raised by mounting evidence that they also permeate groundwater systems. Research on groundwater contamination has traditionally concentrated on chemical contaminants such as pesticides, petroleum hydrocarbons, nitrates, and heavy metals (Lee et al., 2024). But the recent finding of MPs in groundwater has changed the landscape of contamination and presented special difficulties and complications for environmental management. MPs enter groundwater through a variety of interrelated entry points. The main ways that MPs enter groundwater systems are by surface runoff, agricultural practices (such as the use of fertilizers and compost), atmospheric suspension, and swage effluents. These particles accumulate in underground habitats after traveling great distances and penetrating aquifers. Due to their innate resistance to degradation, MPs may remain in groundwater for longer periods of time after they are present, which could lead to long-term contamination and further exposure (Chia et al., 2022; Lee et al., 2024).

Despite accounting for a substantial part (14%) of the global plastic effluence, research on plastic infectivity of farming soils has only lately begun, in contrast to freshwater and marine habitats. The vertical movement of plastic particles in agricultural soils and the acceleration of pesticide transfer into core groundwater caused by plastic are serious concerns. This topic vitally makes the current understanding of plastic transfer and plastic-insecticide connections in farming soils, discovers prospect research directions, and assesses related analytical challenge in order to determine the hazard of the outsized plastic stock in farming soils for groundwater and domestic water provisions. The use of compost, sewage waste as fertilizer, synthetic mulching and packaging, irrigation by raw waste and surface water, and atmospheric fallout are some of the ways that plastic is introduced into agricultural soils (Wanner, 2021). The main basis of atmospheric fallout is tire abrasion, which discharges a lot of plastic elements into the ambient air. These particles can travel great distances and may end up on agriculture soils. Because plastic is more itinerant, harder to notice, and may move throughout soils into primary aquifers similar to normal colloids (particles <1 μm size), the biodegradation-provoked formation of secondary MPs in farming soils is a key cause for worry (Bläsing & Amelung, 2018; He et al., 2018).

7 Hazards of Plastic Pollution

7.1 Hazards of Plastics on Water Environments

Plastics have a variety of negative effects on aquatic ecosystems, including fisheries (damaged gear, reduced catches), hitchhiking (i.e., attaching to and floating with plastics) of organisms, including invasive species and pathogens, and direct health effects in marine organisms from ingestion or entanglement in litter and fishing gear. According to research on microplastics, marine species that consume them may have a variety of consequences, such as intestinal tract blockage, inflammation, oxidative stress, hormone disruption, reproductive effects, and changes in metabolism and behavior (Bethanie & Håkan, 2019; Vethaak & Leslie, 2016). Recent studies, however, indicate that exposure to smaller, nanoplastic particles increases the likelihood of negative consequences. Although the effects of micro- and nanoplastics on marine habitats at the ecosystem level are still mostly unknown, they could include modifications to food chains and nutrient cycles as well as alterations to the microbial communities that grow on plastics. Current research is dominated by two contrasting views: microplastics clearly affect marine ecosystems, and the risks connected with microplastics have not yet been demonstrated to exist, despite some research suggesting that they may have serious effects. Nonetheless, numerous commonly used chemical additives in plastic items have been discovered in marine environments, and these substances disrupt hormones, cause developmental problems, and disrupt reproduction in a variety of vertebrate species, including fish and marine mammals. The origins of these chemicals in marine environments can be traced to diffuse sources (e.g., wastewater, sewage, atmospheric deposition) or leachates from plastic debris (i.e., chemicals like flame retardants, phthalates, and phenols) may leak out of plastic objects into marine waters, which are caused by the widespread use of both plastics and chemicals worldwide (Bethanie & Håkan, 2019).

7.2 Hazards of Plastics on Humans

Although plastic debris has been highlighted as a potential human health hazard, the direct effects of marine plastics on human health have not been thoroughly demonstrated, and no research specifically look into this topic. For instance, when plastic bags obstruct drainage channels and cause floodwaters to rise or when plastic waste serves as a mosquito breeding ground, plastic items may directly affect people.

Furthermore, it has been demonstrated that bacteria, including possible diseases, inhabit microplastics.

Due to particle toxicity, microplastics may potentially have an impact on human health. They are also becoming more prevalent in human food, notably seafood, and water sources (Bethanie & Håkan, 2019; Rochman et al., 2016). There are health consequences for people who are exposed to the toxins in plastic goods. Diseases and pathologies, such as endocrine disturbance, malignant cells, developmental turmoil, and reproductive oddity, have been linked to chemicals found in plastics (Thompson et al., 2009; Trasande et al., 2015). The most extensively researched effect of plastics on human health is actually chemical exposure. For instance, chemicals and additives found in plastic products that come into contact with food (such as packaging, storage containers, and utensils), toys for kids, and electronics expose people (Bethanie & Håkan, 2019; Groh et al., 2018; Zeng et al., 2016). Microplastics may serve as vectors, introducing toxins from the marine environment into species that humans often eat, according to some data. Microplastics, however, probably have little effect on the buildup of chemicals in the food chain; chemicals in materials that come into touch with food are the main source of human exposure. Additionally, there is worry that long-term, low-level exposures to mixes of these chemicals may result in a variety of pandemic diseases that are difficult to identify or ascribe to a single source, as well as nonlinear consequences (European Food Safety Authority Panel on Contaminants in the Food Chain, 2016).

8 Identification of Water Plastic Pollution

Plastics can be identified and described using a variety of methodologies and procedures. The wide variety of paths and sources makes it difficult to identify secondary microplastics. The identification and detection of microplastics may be hampered if the biofilm is not eliminated from their surface. It is possible to quantify water contamination using elemental, material, and organic approaches. Measurements are made of warmth, turbidity, color, smell, taste, and concentration during the physical testing procedure. The presence of plastic at any level can be identified through physical tests; samples from various sources are compared with standard samples for their physical characteristics; any difference from the ordinary limit indicates the stage of pollution. River water has no tang or smell, but marine water has a tart taste because of the higher absorption of various salts. When evaluating the samples physically, color might also be helpful. While clean water absorbs daylight and is devoid of contaminants, turbidity is the muddiness of water caused by a considerable amount of particles there in the water. As a result, turbid water indicates that the water is contaminated. Chemical techniques analyze the levels of various contaminants in liquid sections using the investigative chemistry principle; they can also identify the kinds and characteristics of these contaminants. Chemicals like manufacturing colors, heavy metals like lead, chromium, or plastic can have an adverse effect on aquatic life or people who drink it if their absorption is too high. The class and cleanliness of the water can be ascertained using a few elemental indicators. Measurements of pH, elemental oxygen demand, and biological oxygen demand are some examples of these markers. Utilizing various flora and fauna and microorganisms as biological markers, the biological approaches track many aspects of a freshwater or maritime ecosystem. If a river can support a wide variety of organisms, the class is probably extremely good; if fish cannot live or thrive there, the quality is very bad. This indicates that many pollutants are present and can have an impact on a broad range of organisms that rely on or live in such water supplies (Muneer et al., 2021; Bhardwaj et al., 2024; Stolte et al., 2015).

9 Recommendations

Although there are many unknowns about what will probably happen if plastic garbage builds up in the ecosystem, it is widely agreed that this kind of material shouldn't be in the oceans and that it could endanger the planetary boundaries if it does. Therefore, the precautionary principle should be applied, just as it is when dealing with other global issues like pollinator decline: that is, even though all the facts are still unknown, the evidence suggests that there may be a serious issue that, if ignored, could have disastrous results. The current methods for the production and use of plastics need to be immediately revised because the production of plastic commodities currently accounts for 6% of world oil output, with that percentage expected to increase to 20% by 2050.

Our knowledge of microplastic pollution in water bodies is lacking in a number of areas, which makes it difficult to create all-encompassing mitigation plans. Standardized and coordinated procedures are required for the sampling, analysis, and quantification of microplastics. Comparisons between research and geographical areas are hampered by the inconsistent methodology. It is unclear what the long-term effects and possible synergistic effects with other contaminants will be. Another poorly understood topic is how microplastics help spread other pollutants and infections throughout water bodies. Gaining more knowledge in this area would help us comprehend the wider environmental effects of microplastic contamination.

Furthermore, greater investigation is needed to clarify the origins, movements, and destiny of microplastics in freshwater ecosystems, with an emphasis on comprehending how various settings affect their distribution and movement. In order to create management plans and policies that effectively reduce the negative effects of microplastics on aquatic ecosystems and human health, these research gaps must be filled. A comprehensive strategy that takes into account the origins of microplastics as well as their effects on ecosystems is needed to mitigate microplastic contamination in water bodies.

It is crucial to establish criteria for assessing the ecological harm that microplastics pose. Researchers think that community involvement, regulation, and the use of biotechnological and engineering technologies can all work together to address the problem of microplastic contamination. By promoting plastic recycling, the use of biodegradable bags and non-plastic resources, and the implementation of awareness campaigns on plastic pollution, both governmental and non-governmental organizations may significantly reduce microplastic pollution (Rhodes, 2018).

There are a number of suggestions to reduce the pollution caused by microplastics. These include

- Minimize single-use plastics: Encouragement and implementation of policies aimed at reducing the production and consumption of single-use plastics are necessary. This entails enacting plastic bag bans or taxes as well as encouraging alternatives like reusable bags and containers. Encouragement should be given to those who use products made from plastic trash.
- Enhance trash management: It is necessary to upgrade the waste management infrastructure to keep plastic debris out of waterways. To lessen the possibility that plastic garbage would end up in aquatic habitats, efficient recycling initiatives, waste collecting systems, and waste disposal facilities should be put in place.
- Encourage sustainable practices: It is important to support industries that use and produce plastics in a sustainable manner. To reduce the production of plastic trash, this entails promoting environmentally friendly packaging, cutting back on superfluous packaging, and implementing the concepts of the circular economy.
- Storm-water management: Better storm-water management is necessary to lessen the movement of microplastics from urban areas to bodies of water. It is recommended that green infrastructure solutions and filtering systems be put in place to catch and stop the flow of plastic particles.
- Increase awareness: To inform the public, businesses, and politicians on the effects of microplastic pollution, governmental organizations can hold awareness campaigns such as conferences and field trips. Raising awareness can result in changes in consumption patterns, support for laws addressing plastic pollution, and more responsible behavior.
- Research and monitoring: To gain a better understanding of the origins, distribution, and impacts of microplastics, invest in research. In order to track the amounts of microplastics in various water bodies and identify emerging hotspots, future study should conduct continuous monitoring of the microplastics in regions such as Africa, Asia, and South America where less data has been published. Authorities everywhere should enact new regulations to ensure that plastic pollution in freshwater environments is regularly monitored.
- New technologies: Innovative and economical methods for removing microplastics from water bodies should be investigated and put into practice. This includes the creation of sophisticated analytical methods for freshwater and marine environments, such as filtration systems and skimmers.
- International cooperation: To reduce or eradicate microplastics from the freshwater environment, businesses, and governmental organizations can collaborate. In order to address the global character of microplastic pollution, it is necessary to promote international cooperation. In order to control the production and

disposal of plastics globally, regulations and legislation must be developed and implemented.
- Cleanup campaigns: It is important to support and take part in community-based campaigns to remove plastic garbage from water bodies. These initiatives have the potential to increase awareness and have a direct impact on lowering environmental microplastic levels.

The ecological unit suffers greatly from the disposable tradition of plastic garbage, primarily in the form of substantial and toxic damage. Because plastic is so common, governments everywhere have created countrywide and global laws and regulations (refer Table 1) to counteract and lessen the current threat (Pilapitiya & Ratnayake, 2024).

10 Conclusions

Although there are many unknowns about what will probably happen if plastic garbage builds up in the ecosystem, it is widely agreed that this kind of material shouldn't be in the oceans and that it could endanger the planetary boundaries if it does. Therefore, the precautionary principle should be applied, just as it is when dealing with other global issues like pollinator decline; that is, even though all the facts are still unknown, the evidence suggests that there may be a serious issue that, if ignored, could have disastrous results. The current methods for the production and use of plastics need to be immediately revised because the production of plastic commodities currently accounts for 6% of world oil output, with that percentage expected to increase to 20% by 2050. Moreover, standardized and coordinated procedures are required for the sampling, analysis, and quantification of microplastics. Comparisons between research and geographical areas are hampered by the inconsistent methodology. It is unclear what the long-term effects and possible synergistic effects with other contaminants will be. Another poorly understood topic is how microplastics help spread other pollutants and infections throughout water bodies. Gaining more knowledge in this area would help us comprehend the wider environmental effects of microplastic contamination. However, it is crucial to establish criteria for assessing the ecological harm that microplastics pose. Researchers think that community involvement, regulation, and the use of biotechnological and engineering technologies can all work together to address the problem of microplastic contamination. By promoting plastic recycling, the use of biodegradable bags and non-plastic resources, and the implementation of awareness campaigns on plastic pollution, both governmental and non-governmental organizations may significantly reduce microplastic pollution.

Table 1 Policies and specialized groups to cut down on plastic wastes

Association	Regulation	Tasks
United Nations Organization	Objectives for sustainable progress	It seeks to stop or lessen marine contamination from land-based activities and certifies the preservation and sustainable utilization of all marine assets
United Nations Environment Programme	Resolution regarding microplastics and maritime plastic debris	Make sure that ocean debris and microplastics are permanently removed. Stop or lessen the use of plastics and microplastics in land-based operations
International Union for Conservation of Nature	Make sure that the ocean's microplastics and litter are permanently removed. Prevent or minimize the use of plastics and microplastics in land-based operations	Using a framework and instruments to evaluate the effectiveness of the involvement, guidance is provided for the detection of plastic leaks
United Nations Convention on the Law of the Sea	The 1982 constitution for the oceans came into effect in 1994	Was an unusual attempt to create a stable human life and regulate the use of marine resources
World Health Organization	Evaluation of plastic's environmental impact and effects on human being's health	Promote closer examination of environmental plastics and their effects on human health. Provide techniques for quantifying the amount of plastic that is anticipated to be removed
European Union	Minimize the environmental impact of specific plastic items	To stop and lessen marine litter from both land and marine sources, make sure that waste management practices are environmentally sustainable
Marine Strategy Framework Directive	Reach a favorable state for the maritime environment in Europe	Intends to preserve Europe's maritime ecosystem while permitting the sea's beneficial usage to continue
Oslo and Paris Conservation Commission	Safeguarding the Northeast Atlantic's marine ecosystem	Provide precise instructions for keeping an eye out for marine trash on beaches
Regional Marine Litter Management Action Plan	Intended to control marine debris	Centered on raising societal awareness and educating people about safe trash disposal and addressing the problem of unlawful dumping on deserted beaches

(continued)

Table 1 (continued)

Association	Regulation	Tasks
Pacific Ocean Pollution Prevention Program	Plans of action and policy for Clear Pacific	A thorough plan outlines the region's priorities for managing trash and pollution by concentrating on various fields
Association of Southeast Asian Nations	Taking action against marine litter in constituent states	Devoted to lowering the release of plastic, improving trash reuse through value chain development, and increasing cleanup and minimizing leakage

References

Agarwal, T., Atray, N., & Sharma, J. G. (2024). A critical examination of advanced approaches in green chemistry: Microbial bioremediation strategies for sustainable mitigation of plastic pollution. *Future Journal of Pharmaceutical Sciences, 10*, 78. https://doi.org/10.1186/s43094-024-00645-x

Ahmed, T., Shahid, M., Azeem, F., Rasul, I., Shah, A. A., Noman, M., Hameed, A., Manzoor, N., Manzoor, I., & Muhammad, S. (2018). Biodegradation of plastics: Current scenario and future prospects for environmental safety. *Environmental Science and Pollution Research, 25*(8), 7287–7298. https://doi.org/10.1007/s11356-018-1234-9

Amelia, T. S. M., Khalik, W. M. A. W. M., Ong, M. C., Shao, Y. T., Pan, H. J., & Bhubalan, K. (2021). Marine microplastics as vectors of major ocean pollutants and its hazards to the marine ecosystem and humans. *Progress in Earth and Planetary Science, 8*, 12. https://doi.org/10.1186/s40645-020-00405-4

Andrady, A. L. (2011). Microplastics in the marine environment. *Marine Pollution Bulletin, 62*(8), 1596–1605. https://doi.org/10.1016/j.marpolbul.2011.05.030

Barboza, L. G. A., Dick, V. A., Lavorante, B. R. B. O., Lundebye, A. K., & Guilhermino, L. (2018). Marine microplastic debris: An emerging issue for food security, food safety and human health. *Marine Pollution Bulletin, 133*, 336–348. https://doi.org/10.1016/j.marpolbul.2018.05.047

Beaumont, N. J., Aanesen, M., Austen, M. C., Börger, T., Clark, J. R., Cole, M., Hooper, T., Lindeque, P. K., Pascoe, C., & Wyles, K. J. (2019). Global ecological, social and economic impacts of marine plastic. *Marine Pollution Bulletin, 142*(January), 189–195. https://doi.org/10.1016/j.marpolbul.2019.03.022

Bethanie, C. A., & Håkan, E. (2019). Marine plastic pollution: Sources, impacts, and policy. *Review of Environmental Economics and Policy, 13*(2), 317–326. https://doi.org/10.1093/reep/rez012

Bhardwaj, L. K., Rath, P., Yadav, P., & Gupta, U. (2024). Microplastic contamination, an emerging threat to the freshwater environment: A systematic review. *Environmental Systems Research, 13*, 8. https://doi.org/10.1186/s40068-024-00338-7

Bläsing, M., & Amelung, W. (2018). Plastics in soil: Analytical methods and possible sources. *Science of the Total Environment, 612*, 422–435. https://doi.org/10.1016/j.scitotenv.2017.08.086

Bråte, I. L. N., Huwer, B., Thomas, K. V., Eidsvoll, D. P., Halsband, C., Almroth, B. C., & Lusher, A. (2017). *Micro and macro-plastics in marine species from Nordic waters*. https://api.semanticscholar.org/CorpusID:134335377

Chia, R. W., Lee, J. Y., Jang, J., Kim, H., & Kwon, K. D. (2022). Soil health and microplastics: A review of the impacts of microplastic contamination on soil properties. *Journal of Soils and Sediments, 22*(10), 2690–2705. https://doi.org/10.1007/s11368-022-03254-4

Dang, T. C. H., Nguyen, D. T., Thai, H., Nguyen, T. C., Tran, T. T. H., Le, V. H., Nguyen, V. H., Tran, X. B., Thao Pham, T. P., Nguyen, T. G., & Nguyen, Q. T. (2018). Plastic degradation by thermophilic Bacillus sp. BCBT21 isolated from composting agricultural residual in Vietnam. *Advances in Natural Sciences: Nanoscience and Nanotechnology, 9*(1), 015014. https://doi.org/10.1088/2043-6254/aaabaf

European Food Safety Authority Panel on Contaminants in the Food Chain. (2016). Presence of microplastics and nanoplastics in food, with particular focus on seafood. *EFSA Journal, 14*(6), e04501. https://doi.org/10.2903/j.efsa.2016.4501

Fayshal, M. A. (2024). Current practices of plastic waste management, environmental impacts, and potential alternatives for reducing pollution and improving management. *Heliyon, 10*(23), e40838. https://doi.org/10.1016/j.heliyon.2024.e40838

Frias, J. P. G. L., & Nash, R. (2019). Microplastics: Finding a consensus on the definition. *Marine Pollution Bulletin, 138*, 145–147. https://doi.org/10.1016/j.marpolbul.2018.11.022

Geyer, R., Jambeck, J. R., & Law, K. L. (2017). Production, use, and fate of all plastics ever made. *Science Advances, 3*(7), e1700782. https://doi.org/10.1126/sciadv.1700782

Godfrey, L. (2019). Waste plastic, the challenge facing developing countries—Ban it, change it, collect it? *Recycling, 4*(1), 3. https://doi.org/10.3390/recycling4010003

Groh, K. J., Backhaus, T., Carney-Almroth, B., Geueke, B., Inostroza, P. A., Lennquist, A., Maffini, M., Leslie, H. A., Slunge, D., Trasande, L., & Warhurst, M. (2018). Chemicals associated with plastic packaging: Inventory and hazards. *PeerJ Preprints, 6*, e27036v1. https://doi.org/10.7287/peerj.preprints.27036v1

Halden, R. U. (2010). Plastics and health risks. *Annual Review of Public Health, 31*(1), 179–194. https://doi.org/10.1146/annurev.publhealth.012809.103714

Haward, M. (2018). Plastic pollution of the world's seas and oceans as a contemporary challenge in ocean governance. *Nature Communications, 9*(1), 9–11. https://doi.org/10.1038/s41467-018-03104-3

He, D., Luo, Y., Lu, S., Liu, M., Song, Y., & Lei, L. (2018). Microplastics in soils: Analytical methods, pollution characteristics and ecological risks. *TrAC Trends in Analytical Chemistry, 109*, 163–172. https://doi.org/10.1016/j.trac.2018.10.006

Ivanova, L., Sokolov, K., & Kharitonova, G. (2018). Plastic pollution tendencies of the Barents Sea and adjacent waters under the climate change. *Arctic and North, 32*(32), 121–145. https://doi.org/10.17238/issn2221-2698.2018.32.121

Ivar doSul, J. A., & Costa, M. F. (2014). The present and future of microplastic pollution in the marine environment. *Environmental Pollution, 185*, 352–364. https://doi.org/10.1016/j.envpol.2013.10.036

Karbalaei, S., Hanachi, P., Walker, T. R., & Cole, M. (2018). Occurrence, sources, human health impacts and mitigation of microplastic pollution. *Environmental Science and Pollution Research, 25*(36), 36046–36063. https://doi.org/10.1007/s11356-018-3508-7

Kibria, M. G., Masuk, N. I., Safayet, R., Nguyen, H. Q., & Mourshed, M. (2023). Plastic waste: Challenges and opportunities to mitigate pollution and effective management. *International Journal of Environmental Research, 17*, 20. https://doi.org/10.1007/s41742-023-00507-z

Kushwaha, M., Shankar, S., Goel, D., Singh, S., Rahul, J., Rachna, K., & Singh, J. (2024). Microplastics pollution in the marine environment: A review of sources, impacts and mitigation. *Marine Pollution Bulletin, 209*(A), 117109. https://doi.org/10.1016/j.marpolbul.2024.117109

Lamb, J. B., Willis, B. L., Fiorenza, E. A., Couch, C. S., Howard, R., Rader, D. N., True, J. D., Kelly, L. A., Ahmad, A., Jompa, J., & Harvell, C. D. (2018). Plastic waste associated with disease on coral reefs. *Science, 359*(6374), 460–462. https://doi.org/10.1126/science.aar3320

Law, K. L. (2017). Plastics in the marine environment. *Annual Review of Marine Science, 9*(1), 205–229. https://doi.org/10.1146/annurev-marine-010816-060409

Lee, J. Y., Cha, J., Ha, K., & Viaroli, S. (2024). Microplastic pollution in groundwater: A systematic review. *Environmental Pollutants and Bioavailability, 36*(1), 2299545. https://doi.org/10.1080/26395940.2023.2299545

Mrowiec, B. (2017). Plastic pollutants in water environment. *Environmental Protection and Natural Resources, 28*(4), 51–55. https://doi.org/10.1515/OsZn-2017-0030

Muneer, F., Azam, M. H., Zubair, M., Farooq, T., Ibrahim, M. Rasul, I., Afzal, M., Ahmad, A., & Nadeem, H. (2021). Remediation of water pollution by plastics. In Inamuddin, M. I. Ahamed, & E. Lichtfouse (Eds.), *Water pollution and remediation: Organic pollutants. Environmental chemistry for a sustainable world*, Vol. 54. Springer, Cham. https://doi.org/10.1007/978-3-030-52395-4_3

Paletta, A., Filho, W. L., Balogun, A. L., Foschi, E., & Bonoli, A. (2019). Barriers and challenges to plastics valorisation in the context of a circular economy: Case studies from Italy. *Journal of Cleaner Production, 241*, 11814. https://doi.org/10.1016/j.jclepro.2019.118149

Pietrelli, L., Pignatti, S., & Fossi, M. C. (2018). Foreword—plastic pollution: A short and impressive story. *Rendiconti Lincei, 29*(4), 803–804. https://doi.org/10.1007/s12210-018-0752-1

Pilapitiya, P. G. C. N. T., & Ratnayake, A. S. (2024). The world of plastic waste: A review. *Cleaner Materials, 11*, 100220. https://doi.org/10.1016/j.clema.2024.100220

Pinto, R. B., van Emmerik, T. H. M., Duah, K., van der Ploeg, M., & Uijlenhoet, R. (2024). Mismanaged plastic waste as a predictor for river plastic pollution. *Science of the Total Environment, 951*, 175463. https://doi.org/10.1016/j.scitotenv.2024.175463

Ragossnig, A. M., & Agamuthu, P. (2021). Plastic waste: Challenges and opportunities. *Waste Management & Research, 39*(5), 629–630. https://doi.org/10.1177/0734242X211013428

Rhodes, C. J., Ivanova, L., Sokolov, K., Kharitonova, G., Kondori, B., Mahmudi, R. et al. (2018). The ocean plastic pollution challenge: Towards solutions in the UK. *Nature Communications, 103*(1), 1–4. https://doi.org/10.1038/s41467-018-03104-3

Rhodes, C. J. (2018). Plastic pollution and potential solutions. *Science Progress, 101*(3), 207–260. https://doi.org/10.3184/003685018X15294876706211

Rochman, C. M., Browne, M. A., Underwood, A. J., van Franeker, J. A., Thompson, R. C., & Amaral-Zettler, L. A. (2016). The ecological impacts of marine debris: Unraveling the demonstrated evidence from what is perceived. *Ecology, 97*, 302–312. https://doi.org/10.1890/14-2070.1

Shah, A. A., Hasan, F., Hameed, A., & Ahmed, S. (2008). Biological degradation of plastics: A comprehensive review. *Biotechnology Advances*. https://doi.org/10.1016/j.biotechadv.2007.12.005

Sharma, S., & Chatterjee, S. (2017). Microplastic pollution, a threat to marine ecosystem and human health: A short review. *Environmental Science and Pollution Research*, 21530–21547. https://doi.org/10.1007/s11356-017-9910-8

Sigler, M. (2014). The effects of plastic pollution on aquatic wildlife: Current situations and future solutions. *Water, Air, & Soil Pollution, 225*(11). https://doi.org/10.1007/s11270-014-2184-6

Stolte, A., Forster, S., Gerdts, G., & Schubert, H. (2015). Microplastic concentrations in beach sediments along the German Baltic coast. *Marine Pollution Bulletin, 99*(1–2), 216–229. https://doi.org/10.1016/j.marpolbul.2015.07.022

Strokal, M., Vriend, P., Bak, M. P., Kroeze, C., van Wijnen, J., & van Emmerik, T. (2023). River export of macro- and microplastics to seas by sources worldwide. *Nature Communications, 14*, 4842. https://doi.org/10.1038/s41467-023-40501-9

Thevenon, F., & Carroll, C. (2015). *Plastic debris in the ocean: The characterization of marine plastics and their environmental impacts, situation analysis report*. IUCN. https://doi.org/10.2305/iucn.ch.2014.03.en

Thompson, R. C., Moore, C. J., vom Saal, F. S., & Swan, S. H. (2009). Plastics, the environment and human health: Current consensus and future trends. *Philosophical Transactions of the Royal Society B: Biological Sciences, 364*, 2153–2166. https://doi.org/10.1098/rstb.2009.0053

Trasande, L., Zoeller, R.T., Hass, U., Kortenkamp, A., Grandjean, P., Myers, J. P., DiGangi, J., Bellanger, M., Hauser, R., Legler, J., & Skakkebaek, N. E. (2015). Estimating burden and disease costs of exposure to endocrine disrupting chemicals in the European Union. *Journal of Clinical Endocrinology & Metabolism, 100*, 1245–55. https://doi.org/10.1210/jc.2014-4324

Van Emmerik, T., & Schwarz, A. (2020). Plastic debris in rivers. *Wires Water, 7*, e1398. https://doi.org/10.1002/wat2.1398

Vethaak, A. D., & Leslie, H. A. (2016). Plastic debris is a human health issue. *Environmental Science & Technology, 50*, 6825–6826. https://doi.org/10.1021/acs.est.6b02569

Villarrubia-Gómez, P., Cornell, S. E., & Fabres, J. (2018). Marine plastic pollution as a planetary boundary threat—The drifting piece in the sustainability puzzle. *Marine Policy, 96*, 213–220. https://doi.org/10.1016/j.marpol.2017.11.035

Wang, J., Zheng, L., & Li, J. (2018). A critical review on the sources and instruments of marine microplastics and prospects on the relevant management in China. *Waste Management and Research, 36*(10), 898–911. https://doi.org/10.1177/0734242X18793504

Wanner, P. (2021). Plastic in agricultural soils—A global risk for groundwater systems and drinking water supplies?—A review. *Chemosphere, 264*(1), 128453. https://doi.org/10.1016/j.chemosphere.2020.128453

Yu, Y., Zhou, D., Li, Z., & Zhu, C. (2018). Advancement and challenges of microplastic pollution in the aquatic environment: A review. *Water, Air, and Soil Pollution, 229*(5), 2–18. https://doi.org/10.1007/s11270-018-3788-z

Zeng, X., Xu, X., Boezen, H. M., & Huo, X. (2016). Children with health impairments by heavy metals in an e-waste recycling area. *Chemosphere, 148*, 408–415. https://doi.org/10.1016/j.chemosphere.2015.10.078

Assessing Different Tools Employed in Auto-segregation of Plastic Waste

Pankaj Kumar, Surya Nikhil Mallampalli, Pavan Rahul Konathala, Pooja Vishali Vasireddy, Manasa Mugdha, Hephzi Jones, Priyanka Singh, and Pankaj Pathak

Abstract Segregation and recycling of waste have been recognized to be vital for both economic and ecological reasons, with industries demanding high efficiency. However, current studies on automatic waste detection lack benchmarks and widely accepted standards, making comparisons difficult. This chapter addresses the issue by employing deep learning for waste classification into two categories: recyclable and non-recyclable materials. The dataset explained in this chapter has been compiled from various sources, ensuring a diverse representation of waste types. The garbage classification model is trained on the MobileNetV2 deep neural network architecture, enabling rapid and accurate classification of domestic waste. The model achieved an impressive 94% absolute accuracy, in 15 epochs, showcasing its efficiency and effectiveness. The applications of this research aim to provide better waste categorization and encourage more widespread recycling practices. By leveraging deep learning techniques, the proposed model streamlines the waste-sorting process, potentially saving substantial labor, material, and time costs associated with manual sorting. The development of efficient and accurate waste classification systems can significantly contribute to environmental sustainability efforts and promote a circular economy. The proposed waste classification model automates plastic segregation by making the process more efficient and faster.

Keywords Artificial intelligence · Plastics · Waste segregation · Waste management · Environmental pollution

---

P. Kumar · H. Jones · P. Pathak
Resource Management Lab, Department of Environmental Science and Engineering, SRM University AP, Amaravati, Andhra Pradesh, India

S. N. Mallampalli · P. R. Konathala · P. V. Vasireddy · M. Mugdha · P. Singh (✉)
Department of Computer Science and Engineering, SRM University - AP, Amaravati, Andhra Pradesh, India
e-mail: priyanka.s@srmap.edu.in

P. Pathak
Waste to Wealth Group, SRM University - AP, Amaravati, Andhra Pradesh, India

P. Pathak et al. (eds.), *Plastic Footprint*, https://doi.org/10.1007/978-981-96-6362-0_8

1 Introduction

Solid waste management has become a critical concern globally and faces a significant environmental and economic challenge. India generates a staggering 62 million tons of solid waste annually, ranking it as the world's third-largest waste producer (Pujara et al., 2023). This issue is further compounded by rapid urbanization, population growth, and changing consumption patterns. The annual quantity of solid waste in Indian cities has grown exponentially, from 6 million tons in 1947 to 48 million tons in 1997, with an expected rise to 300 million tons by 2047. This alarming trend stems from a population explosion and a growing preference for packaged goods, leading to an increase in non-biodegradable waste such as plastics. It reveals that non-recyclable materials comprise 43% of food waste compared to only 53% of recyclable materials (Samion et al., 2018). This underscores the urgent need for efficient waste management strategies, particularly those which promote recycling. However, waste segregation is a major hurdle in achieving efficient waste management (Jones & Pathak, 2025). Manual sorting exposes workers to health hazards, and public awareness regarding proper waste categorization remains limited. Many individuals struggle to distinguish between recyclable and non-recyclable items beyond common materials such as paper, plastic, cardboard, and cans. To address these challenges and automate the waste segregation process, we propose a deep-learning-based solution utilizing convolutional neural networks (CNNs). This approach offers a promising avenue for tackling waste management issues in India.

Numerous research efforts have been dedicated to the study of image classification, surrounding feature extraction approaches, and classification algorithms for images (Kumar et al., 2024). Image classification technology harnesses computers to mimic human capabilities in categorizing images according to predefined rules, finding applications across diverse domains such as medical, agricultural, industrial, and service domains. The CNNs have emerged as a widely adopted architecture for deep learning in image processing (Indolia et al., 2018). CNNs require fewer training parameters and effectively capture neighboring information when extracting features from images has rendered them proficient at automatically extracting diverse features during image processing. As a prominent branch of artificial neural networks, deep learning has undergone rapid advancements, leading to its widespread adoption across various industries, including language assistants, autonomous driving, and face recognition (Dhiman et al., 2021). In this study, the focus is primarily on reviewing the applications of computer vision. CNNs are class of neural networks commonly employed for image analysis, which have found applications in image classification, object detection, semantic and instance segmentation, generative adversarial networks, autoencoders, and many more. Given their extensive applications in image and video processing domains, CNNs are well suited for image classification tasks such as categorizing waste materials, hence the suitable approach for this research.

Several studies have explored the application of CNNs for waste material classification. The Trash Net model, proposed by Gary Thung and Mindy Yang (Mittal et al., 2016), categorizes waste materials into six distinct categories, while Compost Net focuses on classifying food waste and compostable waste into seven categories. The Spot-Garbage mobile application employs a CNN architecture called Garb Net to detect and localize garbage in images captured in Indian urban areas, even without categorizing the waste. Furthermore, the model presented by Shi et al. (2021) analyzes the Trash Net dataset for waste image classification. To enhance detection accuracy, a simple and efficient waste classification method with fewer parameters has been proposed, leveraging a multi-layer hybrid CNN, achieving an accuracy of up to 92%. The study presents the basic model along with four successive improved versions, facilitating a comprehensive analysis of classification performance and accuracy. Additionally, adequate heat map comparisons in the form of images are provided to aid rich feature extraction and a comparison of the proposed scheme with existing ones is presented in terms of accuracy, parameters, and complexity.

2 Model and Methodology

This section is structured in seven sections, detailing the research methodology followed in this study. This includes explanations of the dataset description, MobileNetV2, architecture, the waste classification model, loss function, the results achieved, and further a discussion that presents the key findings from the experiments along with relevant conclusions.

2.1 Dataset Description

The dataset used in this study included a variety of plastic and non-plastic items that were carefully chosen to reflect common things encountered in everyday situations as shown in Fig. 1. Plastic products include bottles, plastic bags, food containers, packaging materials, and plastic toys, among many others. The non-plastic group included glass bottles, metal cans, paper bags, wooden objects, cloth bags, etc. (Guo & Li, 2020). A total of 2585 images of plastic items and 2578 images of non-plastic items were included in the dataset. These images were organized into separate folders and divided into a training set of 2476 images and a testing set of 204 images, maintaining a split ratio of 70:30. During pre-processing, techniques such as resizing were applied to standardize the resolution of 440 $\times$ 330 pixels to facilitate processing within the neural network architecture (Castagna et al., 2023).

The images were directly passed to the model for training, with no further labeling or annotation required. Cropping techniques were applied as needed to focus on the relevant parts of the objects within the images. Normalization of the data was performed so that a range of [0, 1] could be achieved so that any potential distortions could be eliminated. Before training commenced, a visual inspection of the dataset was conducted by displaying a random selection of 16 images along with their

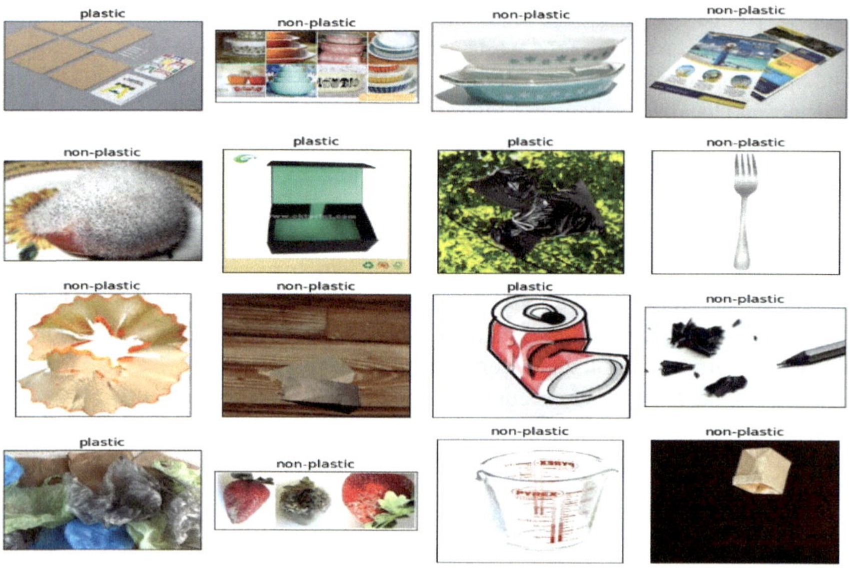

Fig. 1 Sample dataset of images divided into two categories: plastic and non-plastic

corresponding labels. This step ensured the integrity of the dataset and provided insights into the distribution and characteristics of the images. Overall, the dataset was diligently prepared to ensure diversity, balance, and suitability for training the model to accurately detect and classify plastic and non-plastic items. The inclusion of various types of plastic and non-plastic items ensures that the model could be capable of effectively distinguishing between the two categories in real-world scenarios.

2.2 MobileNetV2 Architecture

The architecture, MobileNetV2 as illustrated in Fig. 2, starts with a fully convolutional layer containing 32 filters, followed by a sequence of 19 residual bottleneck layers as described by Zhou et al. To enhance robustness during low-precision calculations, ReLU6 non-linearity function was employed (Broek et al., 1998). Consistent with contemporary network practices, kernel size of 3×3 was utilized throughout the model. Additionally, dropout and batch normalization techniques are incorporated during the training phase. Notably, a constant expansion rate is applied across the network, except the initial layer. From empirical findings of Sandler et al. (2018), it was suggested that expansion rates between 5 and 10 yields are comparable performance curves. However, smaller networks benefit marginally from slightly lower expansion rates, while larger networks exhibit a slight performance boost with higher rates.

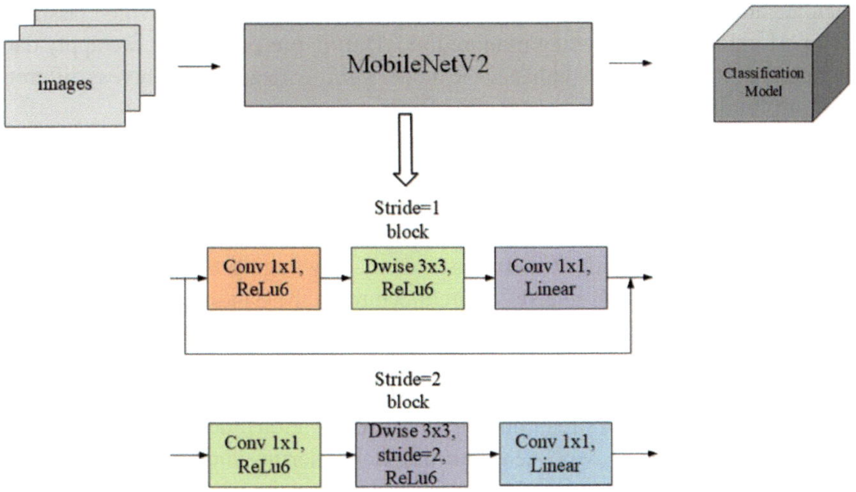

Fig. 2 MobileNetV2 network structure

The inverted residual structure deviates from the standard residual structure in its processing order. It commences with a convolutional downscaling operation, followed by a depth-wise convolution, and concludes with a convolutional upscaling step. Conversely, the traditional residual structure initiates with convolutional upscaling, proceeding with a depth-wise convolution, and finalizes with a convolutional downscaling step. The cornerstone of this network lies in its utilization of depth-wise separable convolutions. This specific convolution method offers significant computational cost reduction and parameter minimization compared to traditional convolution approaches. This structural advantage demonstrably enhances the overall performance of this MobileNetV2 model.

2.3 *Waste Classification Model*

This classification model aims to streamline the waste management process by automating the segregation of plastic and non-plastic items (Guo & Li, 2020). The carefully curated dataset comprising two class labels, plastic and non-plastic, is organized into training and testing sets. We ensure a balanced and representative dataset for training this model. The model's depth-wise architecture makes it ideal for deployment on devices with limited processing power. The dataset has been augmented and normalized to enhance model generalization. It is trained with a categorical cross-entropy loss function. Data augmentation techniques, including rotation, flipping, and zooming, are applied to prevent overfitting. A learning rate scheduler is employed to optimize the training process. Transfer learning is also incorporated to regulate and enable the model to utilize pre-learned features, improving classification performance and reducing training time. During pre-processing, we apply data augmentation techniques to enhance the model's ability to generalize across different lighting conditions (Castagna et al., 2023).

Techniques such as flipping, blurring, and contrast adjustment are employed to augment the dataset, while all images are resized to a standardized resolution of 440 × 330 pixels to facilitate processing within the neural network architecture. Our approach involves training a pre-trained MobileNetV2 model to classify whether an image depicts a plastic or non-plastic item. Leveraging the power of transfer learning, the fine-tune MobileNetV2 architecture on the dataset effectively classify waste items. Once trained, the model can be integrated with cameras for real-time plastic prediction, which offers a seamless approach to recycling at segregation units and recycling facilities. The integration of this model with camera systems offers several benefits. Firstly, it streamlines the waste segregation process, reducing manual labor and human error associated with traditional sorting methods. This automation enhances efficiency and throughput in waste management operations. Additionally, by accurately identifying plastic items in real time, this model facilitates the effective recycling of plastics, contributing to environmental sustainability efforts. From a technical perspective, this model's reliance on the MobileNetV2 architecture ensures both accuracy and efficiency in plastic waste classification. MobileNetV2's lightweight design and efficient feature extraction capabilities make it well suited for deployment on resource-constrained devices, such as embedded systems or mobile devices. Furthermore, the use of transfer learning allows us to leverage the knowledge gained from training on large image datasets, resulting in faster convergence and improved performance on the specific task of waste classification. The proposed waste management solution utilizes a camera, waste bins, and a CNN model to detect and classify waste, refer Fig. 3.

The system employs deep learning with CNN for garbage object detection, as illustrated in Fig. 3. The MobileNetV2 model architecture was chosen for its rapid

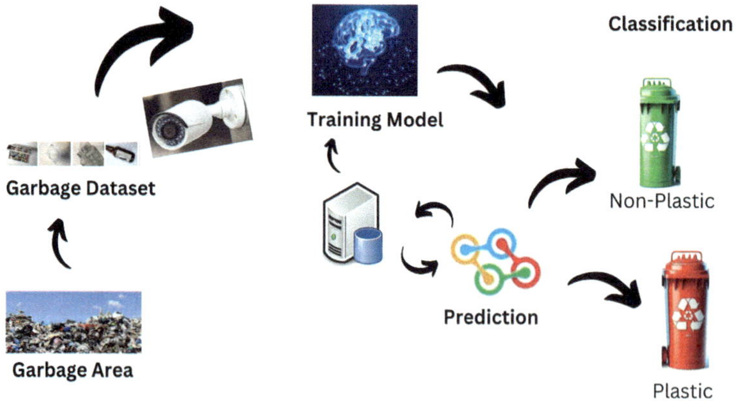

Fig. 3 Proposed architecture of waste classification model

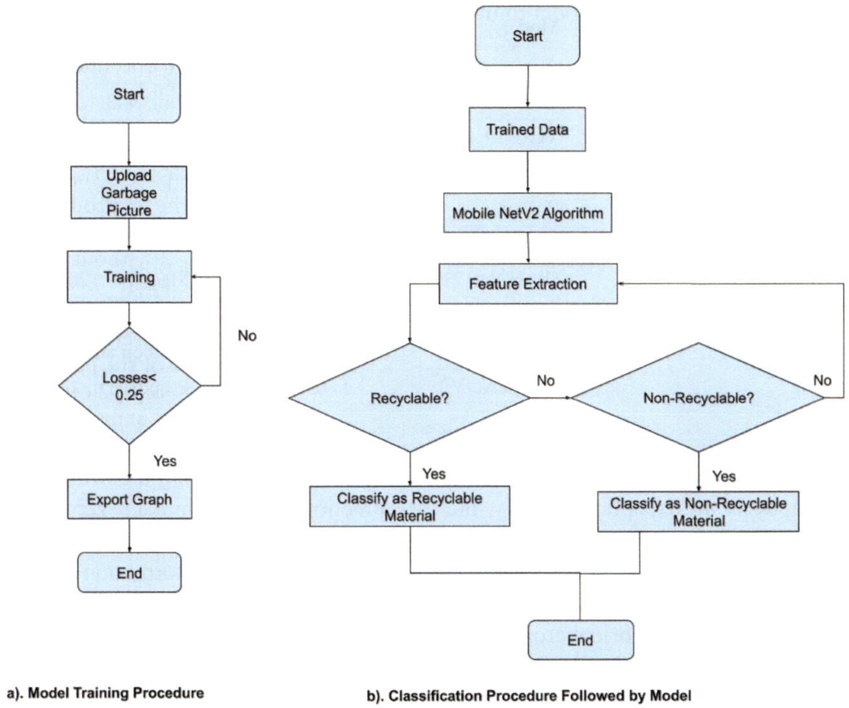

Fig. 4 Algorithms used in automated plastic waste segregation

and highly accurate classification capabilities. The model enables quick detection of multiple objects in an image faster than other models such as the R-CNN series. The MobileNetV2 model also enhances object recognition performance on mobile devices for objects of various sizes. In this study, as shown in Fig. 4, the model classifies whether the input image is plastic or non-plastic based on the training that has been done. Two components are involved in recognition. The first component includes model training where labeled images are inputted. The model's training is halted when the loss function reaches a threshold value of 0.25.

2.4 *Loss Function*

In this study, the performance of the trained model using loss functions is assessed. These functions quantify any discrepancies between the model's predictions and the ground truth values, i.e., the desired outputs (Broek et al., 1998). Higher loss function values indicate greater model inaccuracy. The discrepancy is calculated by comparing the model's outputs on both the training and testing datasets as shown in Table 1.

Table 1 Comparison of the dataset model with the other existing models

Model	Accuracy (%)	Refs.
Autoencoder CNN	90	Ku et al. (2021)
VGG-16	76.94	Ruiz et al. (2019)
VGG-19	79.32	
Inception	87.71	
ResNet	88.66	
Inception ResNet	88.34	
DenseNet121	95	Mao et al. (2021)
MLH-CNN	92.6	Shi et al. (2021)
ResNet-50	87	Adedeji and Wang
MobileNetV2	93.65	Proposed model

Minimizing the loss function is crucial for all machine learning algorithms and this study is no exception. Ideally, the loss function should approach zero under optimal training conditions. In this work, the cross-entropy loss function for classification tasks is employed. Cross entropy, a measure derived from information entropy, quantifies the divergence between two probability distributions (Singh et al., 2019). Specifically, the categorical cross-entropy function is utilized suitable for multiclass classification problems, where each data point is assigned to a single class. For instance, materials like plastics, papers, and paper bags belong to the "recyclable" class, while organic waste such as fruits, vegetables, ceramics, and kitchenware fall under the "non-recyclable" category as shown in Fig. 5. Equation (1) can determine the cross-entropy classification loss value:

$$\mathrm{CE} = -\sum_{i=1}^{C} t_i \log s_i \tag{1}$$

3 Result and Discussions

In this section, a comprehensive evaluation of the proposed method for plastic detection is presented. This includes detailed graphs illustrating the loss and accuracy metrics across different epochs, which provide insights into the model's training progress and performance. Additionally, the confusion matrix is analyzed to understand the classification accuracy and identify any misclassifications between plastic and non-plastic items. The evaluation is subdivided into several key parts: an in-depth analysis of the classification loss function, an examination of the smooth L1 loss in the regression component, and a thorough discussion of the detection and classification results. These evaluations help to validate the model. Below shown is the collage of images, which were labeled by the model, and each image shown in

Fig. 5 Dataset images with their predicted and original class labels

Fig. 6 has two labels: one is original and another is predicted (Castagna et al., 2023). The progressive reduction in loss, presented in Fig. 6, with each epoch indicates the model's accurate learning process, demonstrating its potential effectiveness with real-time data. As the training progresses, the consistent decrease in loss, approaching near-zero values, affirms the model's high precision and readiness for deployment in real-world scenarios. This steady improvement underscores the efficacy of the model and its capability to perform reliably in practical applications of plastic detection.

The trained model has achieved 94% accuracy on test data. A consistent increase in accuracy was observed with each epoch, reflecting the model's improving performance over time. This steady rise in accuracy highlights the model's effective learning process and its growing ability to correctly classify plastic and non-plastic items. Such improvement suggests that the model is becoming more adept at distinguishing between different types of waste, further supporting its readiness for deployment in real-world applications.

The confusion matrix reveals that the model achieved 676 true positives and 460 true negatives, demonstrating its strong capability in accurately identifying both plastic and non-plastic items. The remaining 77 instances represent misclassifications, which are relatively few compared to the correctly identified cases. This distribution underscores the model's high precision (~94%) and effectiveness in distinguishing between the two categories, indicating its potential for successful deployment in real-world waste management (Shahi et al., 2022).

The results of our study demonstrate the effectiveness of the MobileNetV2 model in accurately classifying plastic and non-plastic waste, with significant improvements in both precision and accuracy as training progressed. The 94% accuracy and fine

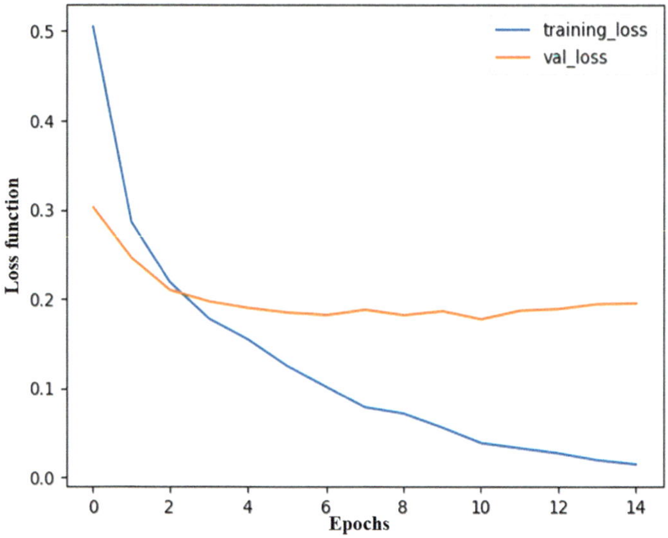

Fig. 6 Loss curve MobileNetV2 model on trained dataset of original and predicted data

precision was achieved. The confusion matrix highlights a strong performance. These findings indicate the model's virtue and its potential for incorporation into daily life and real-time usage (Ramos et al., 2024).

As compared to previous studies, this model exhibits faster and more accurate detection capabilities on plastic wastes, particularly due to the use of MobileNetV2 architecture, which is optimized for mobile as well as embedded device performance. This advantage makes this solution particularly suitable for deployment in resource-constrained and household environments. The implications of our results are substantial, suggesting that this approach can significantly improve the efficiency of waste-sorting processes by accurately identifying and classifying waste types. This model can contribute to more effective recycling and waste management strategies, ultimately leading to reduced environmental impact. Currently, the model is customized to take input from a specified folder path and label it accordingly. However, this study has limitations. The model's performance could be further validated with a larger and more diverse dataset to ensure generalizability across different waste types and conditions. Additionally, the system's real-time deployment in dynamic environments needs further development. Future work puts light on enhancing the model's accuracy and scale, possibly through the integration of additional data augmentation techniques or the incorporation of more advanced neural network architectures. Moreover, exploring the use of this model in other applications, such as sorting industrial waste or e-waste, by integrating existing or new approaches.

4 Conclusions

This study investigates the potential of convolutional neural networks (CNNs) for rapid and accurate garbage classification methods. The proposed system, employing the infamous MobileNetV2 architecture, successfully identifies and categorizes two types of waste: recyclable and non-recyclable materials, while achieving an accuracy exceeding 90%, which is remarkable. Nevertheless, there is always room for potential advancements. These advancements hold promise for enhancing the overall performance of the system. Furthermore, this project emphasizes the potential of CNNs in revolutionizing waste management practices. The achieved accuracy of 93.65% in identifying the tested garbage types demonstrates the effectiveness of this approach. Building upon this groundwork, future studies can explore the integration of a robotic arm for waste separation based on CNN's classifications. Such advancements can pave the way for a more automated and environmentally friendly waste management system.

References

Castagna, A., Dierssen, H. M., Devriese, L. I., Everaert, G., Knaeps, E., & Sterckx, S. (2023). Evaluation of historic and new detection algorithms for different types of plastics over land and water from hyperspectral data and imagery. *Remote Sensing of Environment, 298*, 113834. https://doi.org/10.1016/j.rse.2023.113834

Dhiman, S., Srivatsan, K., & Jain, A. (2021). Waste classification using transfer learning with convolutional neural networks. *IOP Conference Series: Earth and Environmental Science, 775*(1). https://doi.org/10.1088/1755-1315/775/1/012010

Guo, X., & Li, P. (2020). Mapping plastic materials in an urban area: Development of the normalized difference plastic index using WorldView-3 super spectral data. *ISPRS Journal of Photogrammetry and Remote Sensing, 169*, 214–226. https://doi.org/10.1016/j.isprsjprs.2020.09.003

Indolia, S., Goswami, A. K., Mishra, S. P., & Asopa, P. (2018). Conceptual understanding of convolutional neural network—A deep learning approach. *Procedia Computer Science, 132*, 679–688. https://doi.org/10.1016/j.procs.2018.05.069

Jones, K. H., & Pathak, P. (2025). Recycling technologies and resource recovery from plastic waste. In *Valorization of resources from urban mined materials world scientific series on advances in environmental pollution management* (pp. 61–80). https://doi.org/10.1142/9789811297755_0004

Ku, Y., Yang, J., Fang, H., Xiao, W., & Zhuang, J. (2021). Deep learning of grasping detection for a robot used in sorting construction and demolition waste. *Journal of Material Cycles and Waste Management, 23*(1), 84–95. https://doi.org/10.1007/s10163-020-01098-z

Kumar, P., Jones, H., Pathak, P., & Singh, P. (2024). Automated segregation of plastic wastes: Pioneering a greener frontier in waste management. In P. Pathak, S. Ilyas, R. R. Srivastava, J. Dar, & S. Kothandaraman (Eds.), *Advances in environmental sustainability, energy and earth science*. AESEE 2024. Springer, Cham. https://doi.org/10.1007/978-3-031-73820-3_1

Mao, W. L., Chen, W. C., Wang, C. T., & Lin, Y. H. (2021). Recycling waste classification using optimized convolutional neural network. *Resources, Conservation and Recycling, 164*. https://doi.org/10.1016/j.resconrec.2020.105132

Mittal, G. Y., Garg, K. B., & Krishnan, N. C. (2016). Spot Garbage: smartphone app to detect garbage using deep learning. In *UbiComp 2016—ACM international joint conference on pervasive and ubiquitous computing* (pp. 940–945). https://doi.org/10.1145/2971648.2971731

Pujara, Y., Govani, J., Patel, H. T., Pathak, P., Mashru, D., & Ganesh, P. S. (2023). Quantification of environmental impacts associated with municipal solid waste management in Rajkot city, India using life cycle assessment. *Environmental Advances, 12*, 100364. https://doi.org/10.1016/j.envadv.2023.100364

Ramos, E., Lopes, A. G., & Mendonça, F. (2024). Application of machine learning in plastic waste detection and classification: a systematic review. *Processes, 12*(8), 1632. https://doi.org/10.3390/pr12081632

Ruiz, V., Sánchez, Á., Vélez, J. F., & Raducanu, B. (2019). Automatic image-based waste classification. In *Lecture notes in computer science (including subseries lecture notes in artificial intelligence and lecture notes in bioinformatics)* (Vol. 11487, pp. 422–431). https://doi.org/10.1007/978-3-030-19651-6_41

Samion, N. K., Faezah Adan, N. M., & Kadir, A. (2018). The effectiveness of segregation recyclable materials by automated Motorized Bin. ideCON 2289-8107. https://www.researchgate.net/publication/325986121

Sandler, M., Howard, A., Zhu, M., Zhmoginov, A., & Chen, L. C. (2018). MobileNetV2: Inverted residuals and linear bottlenecks. In *Proceedings of the IEEE computer society conference on computer vision and pattern recognition* (pp. 4510–4520). https://doi.org/10.1109/CVPR.2018.00474

Shahi, T. B., Sitaula, C., Neupane, A., & Guo, W. (2022). Fruit classification using attention-based MobileNetV2 for industrial applications. *PLoS ONE, 17*(2), e0264586. https://doi.org/10.1371/journal.pone.0264586

Shi, C., Tan, C., Wang, T., & Wang, L. (2021). A waste classification method based on a multilayer hybrid convolution neural network. *Applied Sciences, 11*(18). https://doi.org/10.3390/app11188572

Singh, B., Toshniwal, D., & Allur, S. K. (2019). Shunt connection: An intelligent skipping of contiguous blocks for optimizing MobileNet-V2. *Neural Networks, 118*, 192–203. https://doi.org/10.1016/j.neunet.2019.05.024

Van Den Broek, W. H. A. M., Wienke, D., Melssen, W. J., & Buydens, L. M. C. (1998). Plastic material identification with spectroscopic near-infrared imaging and artificial neural networks. *Analytica Chimica Acta, 361*(1–2), 161–176. https://doi.org/10.1016/S0003-2670(98)00084-8